# Keep It Rolling

*By*
*Vincent Buggs*

# Table of Contents

# Dedication

This book is a heartfelt tribute to all who serve and to those who have made the ultimate sacrifice.

To my mother, Jessica, an extraordinary individual whose remarkable resilience allows her to sail through life's stormy seas, rising above challenges when others might falter. She has always been the unwavering rock of our family, instilling in us a fundamental rule: never let the world define your worth; instead, demonstrate your value to the world. As a child of the civil rights movement, she witnessed America's struggle for humanity and emerged as a beacon of hope. She embodies the belief that education and hard work unlock the doors to opportunity.

To my father, Senior Master Sergeant Alton G. Buggs, Air Force, a man of few words Rest in Peace, Dad. My stepfather, Master Sergeant A.B. Spencer of the United States Air Force, exemplified an extraordinary work ethic, dedicating himself tirelessly to both service and family.

In every challenge I encounter, my wife Cheryl stands steadfast by my side not merely as a loving partner, but as my battle buddy and greatest source of inspiration. Our children and grandchildren are the vibrant driving force behind my unwavering commitment to uplift others, fueling my desire to make a positive impact in their lives.

I want to express my deepest gratitude to the extraordinary individuals whose unwavering support and wisdom have profoundly shaped my journey. First and foremost, I am especially thankful to Mrs. Helen Jackie Yates, a remarkable mother of students of Georgia Southern University. Her nurturing spirit and constant encouragement have served as a guiding light in my life, providing me with both emotional strength and inspiration during challenging times. Her belief in my potential has motivated me to push beyond my limits and strive for excellence.

Additionally, I owe a significant debt of gratitude to Dr. Charles Thomas, a brilliant professor in the Department of History at Georgia Southern. His insightful teachings and passionate approach to learning have ignited my love for knowledge and fostered my critical thinking skills. Dr. Thomas's dedication to his students and his ability to make complex concepts accessible have not only enriched my academic experience but have also inspired me to pursue my own path as a lifelong learner.

I would like to express my profound gratitude to the remarkable retired Major General Jessica Wright, United States Army Trailblazer, whose exceptional leadership and unwavering strength have inspired me to pursue excellence in all my endeavors relentlessly. I am equally appreciative of retired Colonel D.V. Childs, whose steadfast dedication and unwavering commitment to service have imparted vital lessons about the true essence of integrity. I am also thankful for retired Lieutenant Colonel Wayne Heard, whose profound wisdom and extensive experience have provided me with invaluable insights into resilience and determination. The encouragement and mentorship I have received from these outstanding individuals have been pivotal in shaping my personal and professional growth. I carry their invaluable lessons with me as I navigate my journey forward.

I want to take a moment to express my heartfelt gratitude to my little brother, Ian. From the very first moment you entered this world, we've stood side by side as battle buddies, navigating the ups and downs of life together. Your unwavering spirit and dedication in the military serve as a beacon of inspiration as you carve out your own remarkable path.

I want to take a moment to give a heartfelt shoutout to my extraordinary sisters Fran, Sha-rone, and Candy. You each possess a remarkable gift for infusing every moment we spend together with joy and laughter. As the original pillars of our family, you are the unwavering support system that lifts us all up, providing strength and resilience in every challenge we face. Your presence is a treasured blessing, reminding us of the profound power of unity and the unbreakable bond we share.

Lastly, I want to extend my deepest appreciation to my beloved adopted brother, Noel Mitchell, whose unwavering support and encouragement have been a constant source of inspiration throughout this journey. I also want to express my heartfelt gratitude to my dear friends, Mrs. Lori and Fred Harper. Your exceptional artistic talents, from the beautiful illustrations to the innovative design elements, have truly breathed life into this book. The vibrant colors and thoughtful enhancements have transformed it into something truly spectacular. Thank each of you for your invaluable contributions; this project would not have reached its full potential without your creativity and dedication!

"Keep It Rolling" is a hand-drawn composition shaped in the form of an azimuth—a navigational instrument reimagined as a visual compass of life, leadership, and legacy.

In navigation, an azimuth measures the bearing between one's position and true north; here, the true north star is embodied in a sea of red over one shoulder. Personifying General Buggs's flag, the primary star represents a beacon of moral clarity and steadfast devotion to service, mentorship, and country. Above the opposite shoulder ascends a field of deep blue, carrying seven stars, symbols of unity, endurance, and faith. Stripes of red and white bind personal perseverance to the American narrative of progress and sacrifice.

Anchoring the horizon line, the curved base of the azimuth forms the outline of a slaver—a remembrance of origins and passage. What began as a vessel of suffering now carries collective strength and continuity, representing the ancestral beginning from which generations have risen to leadership. At the heart of the composition, General Buggs stands as the central axis—the point toward which all lines return.

Rounding the foundation of the azimuth are a constellation of symbols that narrate his story: the trucks and their drivers, honoring those whose discipline and excellence sustain the mission. The atomic emblem, expressing raw power, and the military police (MP) officer directing traffic embody command and the proper stewardship of power. Together, the trucks, atom, and MP depict leadership in dynamic balance—authority that channels potential into service, ensuring power flows with purpose, restraint, and responsibility towards accomplishing the mission. Together, the beret and these elements connect modern soldiers to generations before them, linking history, identity, and service.

Right of these forces are symbols of importance and personal connection to the man inside the uniform. Visible identification tags honor fallen comrades whose sacrifices echo through the story of a general's journey. The four young women convey the general's transmission of knowledge and character across generations. Nearby, a cowboy hat evokes the love of country music, while a football highlights a love for coaching and mentorship.

A duality of perpetual leader. At the top of the azimuth are two flags, that of a country (many) and that of a general (one). At the bottom, one type of field (war) transitions to another (life) as the base of the azimuth. Through its measured geometry and layered symbolism, "Keep It Rolling" functions as both compass and chronicle—a visual orientation of life, leadership, and legacy.



**NOel SAint MItchell**

# Forward

**"Your only obligation in any lifetime is to be true to yourself."
James Cromwell**

As we journey through our formative years, many of us are filled with vibrant dreams and ambitions, painting a picture of what our future might look like. However, as we transition into adulthood, those once-bright aspirations often fade into the background. This phenomenon can largely be attributed to the realities of everyday life.

Life has a remarkable way of carving out our paths, and with each twist and turn, it imparts valuable lessons. I have come to learn that while we each hold individual goals close to our hearts, they can sometimes be overshadowed by the demands and expectations of our circumstances. These external pressures, whether from family, work, or societal norms, often lead us to pause or even abandon our personal dreams for the sake of others.

When we let such forces dictate the direction of our lives, we risk falling into a state of regret and discontent. It's all too easy to feel a sense of yearning for what could have been, if only we had persisted in chasing our goals. Deep down, we all strive to realize the best version of ourselves an ideal that often seems just out of reach. Yet, when faced with various challenges, we tend to compromise our true selves, allowing pieces of our identity to slip away. It's a delicate balance between nurturing our own aspirations and attending to the needs of those around us, and often, it is that very struggle that shapes the fabric of our lives.

There are moments in life when you must step back from those closest to you to explore a different path and discover a new way of living. After graduating from college, I made the difficult decision to create some distance from my mother and sisters for several years. This was a time when I felt an urgent need to concentrate on my aspirations and the goals I had set for myself. During this period of separation, I began to recognize a profound truth: my mother, despite her love and support, did not fully grasp the intricate, often challenging landscape young men must navigate on their journey to adulthood.

The transition into manhood can present a series of trials that can be relentless and, at times, harsh. This realization prompted me to reflect on the essential role that a father figure plays in shaping a boy's life. Fathers, in many ways, act as beacons of guidance, offering not just encouragement but also instilling a sense of respect and, at times, fear. It was only after I embraced the role of a fatherhood myself that I fully appreciated the impact I could have on my children's lives.

As a parent, I came to understand that children thrive in environments filled with positive role models, and it is crucial for them to be surrounded by strong male figures who can provide support and guidance. I consistently emphasize this message in my writing because I wholeheartedly believe that, for children to grow into well-rounded individuals capable of achieving success, they need a nurturing, balanced environment that fosters their development.

Life requires common sense and the ability to make good decisions in complex situations. Dissecting these situations begins the moment a child learns to stand up. Throughout life, we will have to rise to our feet countless times.

During my time in college, I met Dr. Charles Thomas, who became a father figure and mentor to me. He took the time to speak to me in a way that no one else ever had. He helped me prioritize my life and understand the importance of staying in the military. This prompted me to rethink my life and discover a new challenge to tackle.

My experiences with pain, heartache, and depression shaped my goals and values. I stopped pretending and began to live truly. I started to realize the value of time and the importance of the people around me. The time you invest in good people can change your life. Dr. Thomas taught me an important lesson: I can't change my race or the color of my skin, but I can transform my environment. This is a life lesson I never forget.

# Preface:
# Keep It Rolling!!

The heaviest weights we carry in life aren't forged from steel or iron but rather consist of the unmade decisions we hold onto, often without realizing their burden. Each day presents us with a myriad of choices, and these choices can dramatically alter the course of our lives. We must strive to create a life rich with purpose and fulfillment. We must avoid letting our fleeting emotions dictate the outcomes of these choices. Instead, we should cultivate a structured approach to decision-making, guided by a clear process and well-defined plans. By fostering a deep and supportive relationship with ourselves, we can initiate profound transformations in our lives.

In my reflections on the journey titled "Keep It Rolling," I candidly confront my own shortcomings and the myriad challenges I have faced along the way. No amount of advice from self-help literature can substitute for the active decision-making that everyone must engage in to enact real change. The choice to move forward rests solely on our shoulders. Eventually, we all reach pivotal moments in our lives when we must consciously decide to leave the weight of our past behind and wholeheartedly embrace the future ahead of us. I encountered such a moment, a turning point, where I had to gather my resolve and remind myself to keep it rolling! This mantra became a beacon of hope, encouraging me to push past obstacles and keep striving for a brighter tomorrow.

# Introduction

I would like to thank you for purchasing my book and for taking the challenge to transform your life. Reaching any goal begins with your decision to make changes. The key to living a fulfilling life lies in recognizing that you can make a difference by choosing the path you want to follow. The greatest struggle we face is admitting that we need to create change. Too often, we find ourselves merely existing instead of truly living with purpose.

In 2020, I faced an unprecedented challenge that tested my resilience and character like never before. For the first time in my life, I found myself subjected to unfounded personal attacks that struck deeply at my core, conflicting with my values and moral beliefs. It was a disorienting experience that forced me to confront uncomfortable truths about human behavior and societal dynamics.

This moment felt like a leap into the unknown, where every step forward was fraught with uncertainty. Like many, I have always encountered chaos in various forms however, this episode was particularly tumultuous, as it came from unexpected sources. It served as a pivotal turning point for me and compelled me to reevaluate my priorities and the direction of my life.

Determined not to succumb to the debilitating traps of confirmation bias or the insidious nature of microaggressions, I made a conscious choice to maintain my integrity. Instead of retaliating or losing myself in the chaos, I embraced the opportunity for personal growth and self-reflection. I sought out supportive networks and engaged in open dialogue to better understand others' perspectives.

Throughout this journey, I held firmly to the belief that my future was in my hands. I was resolute in ensuring that my narrative would be authentically written by me, not defined by the negativity surrounding me. This experience not only shaped my outlook on adversity but also ignited a passion within me to advocate for authenticity and understanding in every interaction.

**"Keep It Rolling" is about human awareness and discovering your inner self. While motivation can be fleeting, inner strength endures forever. Life is a series of ups and downs, and I have experienced my share of failures. This prompted me to create the MSD (Mission Strategy and Destination) framework. After leaving the military, I started my own business despite having no prior experience in entrepreneurship. What I did have, however, was over 30 years of experience with creative and visionary principles that I had honed while coaching and mentoring others.**

**The critical juncture in an individual's life occurs when one begins to adhere to their own counsel. How can we effect meaningful change in our lives? It is achieved through straightforward decisions taken progressively, one day at a time.**

I firmly believe that there is immense value in allowing others to perceive themselves as victorious. This approach creates a golden opportunity to surprise them, a lesson I learned from my mother. When doubt casts its shadow over me, I feel a surge of strength and gain deeper insights into my character. Each failure becomes a steppingstone, rich with lessons, guiding me toward new and uncharted chapters in my life.

Pursuing greatness is not just a goal for me; it is a daily commitment, a relentless journey. There are moments when discouragement creeps in like a persistent fog, yet I stand firm in my resolve to make progress. Every day, I am propelled to engage in activities that may not always ignite my passion. I am inspired to complete these activities by the tireless efforts and unwavering spirit of a few extraordinary individuals of color. Their tireless efforts and unwavering spirit, often in the face of adversity, continue to light my path and inspire me to strive for a world abundant with opportunities.

I would like to share my perspective regarding a significant aspect of my heritage. My ancestry includes individuals who endured the hardships of slavery and were stripped of their dignity and cultural identity. While this statement may not align with the prevailing notions of political correctness in various contexts, it is crucial for understanding my viewpoint within the current framework of Western society. Although we all face daily challenges, none can truly compare to the profound struggles my predecessors faced.

I entered this world in a modest home nestled just off a tar road, welcomed into life by a midwife named Mrs. Maddy. Even now, I find myself in awe of the remarkable journey that is human birth: how we all emerge, gasping for air, as our stories begin to unfold. In those early years, the socioeconomic divide that usually shapes our lives seems to vanish, creating a rare, level playing field. For the first four years, we all confront the same monumental challenge: the struggle to stand on our wobbly legs, with our sights set on the thrilling promise of running freely one day. It may seem like a familiar sentiment, but it perfectly captures the essence of our human resilience, a testament to the fervent spirit that unites us all.

I've often found myself pondering why so many people grapple with disappointment as they age. For me, disappointments haven't just been bumps in the road; they've been invaluable moments that have shaped my worldview and strengthened my determination to tackle life's hurdles. Life isn't just a series of big, sweeping choices; life is about those small, intentional decisions that pave the way for meaningful opportunities. To reach a significant goal, it's essential to embrace the power of consistent, incremental choices that can ultimately lead to remarkable outcomes.

Throughout my journey, I have faced a myriad of challenges and hardships that have shaped a profound resilience within me one that many might struggle to uphold. In the spirit of the ancient Greeks, I see success embodied in the figure of a leader who courageously burns their ships, eliminating any possibility of retreat. Life, ultimately, is a relentless march forward, a quest for growth and understanding, even when confronted with tragedies and setbacks that threaten to derail us.

I spent a considerable amount of my formative years surrounded by people who lacked faith or confidence in me. This is an experience that a lot of individuals encounter at various points in their lives. One of the key lessons I've taken from my journey is that there are often more people willing to revel in your setbacks than those who champion your achievement.

My final assignment in the military imparted invaluable life lessons that exceeded any prior experiences. I found myself in a significantly flawed and highly toxic environment, where many individuals had unfortunately become mere shadows of their former selves. I had not witnessed such inadequate leadership since the mid-1990s.

Years before my last assignment, I attended training at the Command and General Staff College at Fort Leavenworth, Kansas. I will always remember the poignant remarks of a senior leader. His opening address began with, 'Instructors, we failed you. He expressed his frustrations regarding past policies and ideologies that contributed to the emergence of risk-averse leaders within the Army. He acknowledged, 'We created these individuals, and now it is your generation that will face the consequences of our failed educational leadership

These statements resonated deeply with me, reflecting my own experiences during my final Army assignment.

I share this event because we all encounter unbelievable moments that lead us to think, "You can't make this stuff up." It's essential to engage wisely; connecting with those who have already made their decisions can often be unproductive. When I chose to leave the Army, I embraced my own path, undeterred by external opinions. Getting into my BMW Z4 and driving back to Florida was a profound moment of relief and liberation. At some point in your life, you take the option out of other people's hands and Keep it Rolling!

**"Leave nothing for tomorrow which can be done today." George Washington.**

# Chapter 1:
# No Time for Heroes

Success often emerges from an uncomfortable journey that requires us to grow daily. My own path began at the tender age of seven when my father, Senior Master Sergeant Alton Buggs (Retired Air Force), returned from Vietnam deeply affected by his experiences and carrying unseen burdens that would shadow our lives. It wasn't until I completed my second tour in Iraq that I truly began to grasp the weight of those hidden struggles. As men, we sometimes overlook the mental scars that are passed down from father to son. My story mirrors many others from the Vietnam era, highlighting the lasting impacts of war on families. As a military brat, I witnessed these wounds up close.

We all bear scars some inflicted by war, others shaped by our childhoods. Yet these scars represent our unique battles. Unfortunately, many individuals never confront these emotional wounds. My father, for instance, kept his own struggles hidden until he learned I would be going into combat. I'll never forget the day he showed up at my house unannounced to share his experiences from Vietnam. This moment was transformative for us both, especially considering my father's naturally introverted nature. He rarely opened up unless discussing chores or corrections. He had been wrestling with his demons from the war while navigating our father-son relationship.

I remember standing in my cozy living room, the afternoon light filtering softly through the curtains, creating a serene backdrop for a conversation that would forever linger in my mind. My father, a pillar of strength in my life, peered down as if searching for the right words, each one weighed with the gravity of his past. He had always kept his experiences in Vietnam shrouded in a veil of silence, rarely sharing the memories that haunted him. But that day was different; the air felt charged with emotion.

As he began to speak, I could see the flicker of distant memories in his eyes. He recounted a harrowing event: the day a truck he drove was obliterated by a hidden sticky bomb. His voice trembled slightly as he recalled the devastation, the chaos that erupted in what felt like mere seconds. At just 19, he had been thrust into the horrors of war, witnessing the fragility of life, realizing how combat could extinguish it in the blink of an eye. In that moment, I understood the weight of his experiences, the burdens he carried, and the lessons embedded within his silence.

He gently reassured me that I could handle whatever challenges lay ahead. "You don't need to be a hero," he said, a soft smile playing on his lips as he shook my hand. That moment remains etched in my memory and offered a rare glimpse of his authentic self. Though he was a man of few words, this brief encounter conveyed a depth of understanding and care that spoke volumes. Not long after, I would be confronted with the harsh and unyielding realities of war.

For the next several months, I would immerse myself in rigorous training and preparation, my mind always anchored by the sobering reality that some of the individuals I would encounter and train alongside might not return from this mission. Each day heightened my awareness that my own time on this earth could also be ending. War has a profound way of illuminating the fragility of life, forcing people to confront the stark truth that time is a finite gift.

In 2004, I was on my first tour in Iraq. As I sat inside the cramped, metallic interior of a C-130 in Kuwait, the stifling heat pressed down on me like a heavy blanket, increasing my anxiety. The air felt thick and stagnant, filled with the distinct scent of fuel and sweat. As I glanced around the dimly lit cabin, I caught a glimpse of the uncertainty etched on my fellow soldiers' faces eyes wide and restless, mirroring my own trepidation. Everyone seemed lost in their thoughts, grappling with the profound transition from the familiarity of home to the daunting unknown that awaited us in a hostile environment. The hum of the aircraft's engines was a constant reminder of our approaching fate, amplifying the tension that hung in the air.

Experiencing challenging environments can fundamentally change our perspective on the world. Many people claim that war transforms individuals, but I believe it reveals our shared humanity. In times of hardship, even the simplest comforts a cold drink of water or a piece of fresh bread become precious treasures. War reshapes our understanding, deepening our appreciation for life and highlighting the beauty and fragility that surround us in our daily lives.

In the months ahead, I would bear witness to the extraordinary courage and boundless compassion of those around me, intertwined with their profound pain and anguish. The fragile thread of life became strikingly apparent amid those harrowing circumstances.

After several months spent in the theater of operations, I found myself on a somber night, the clock striking 2300 hours, aboard a C-130 military aircraft. The dim cabin was filled with an eerie stillness, broken only by the low hum of the engines. My heart felt heavy, burdened by the solemn cargo I accompanied a coffin draped in the vibrant Stars and Stripes, a symbol of sacrifice and honor. As I gazed at the flag fluttering gently despite the confined space, I couldn't help but reflect on the life it represented and the profound loss that weighed on all of us. The sight was heart-wrenching, a poignant reminder of life's fleeting nature. Inside that solemn casket lay a young man, just beginning his journey, barely older than my own father had been during his service in Vietnam. His untimely departure cast a shadow on the hopes and dreams he would never fulfill, a stark testament to the cost of conflict.

We all embark on a journey filled with uncertainty from the moment we enter this world. As the years pass, we inevitably march toward the twilight of our lives. Yet it seems the divine or cosmic order allows us only to glimpse fragments of this intricate blueprint. Standing before the polished casket of the fallen Marine, I couldn't help but feel an overwhelming sadness his mother or father would never again guide him through the ups and downs of life. Instead, he now rests in the care of angels, embraced by eternal peace.

As I navigate the unique challenges of raising my daughters, I am acutely aware of the importance of offering them guidance a gentle compass that will help them face the complexities that lie ahead. I often tell them that, much like a compass that unfailingly points north, they too can find their way through life's uncertainties. If they ever feel lost or unsure, I encourage them to pause, take a deep breath, and recalibrate their course. It's in those moments of reflection and adjustment that they can rediscover their true path, fortified by the love and wisdom that surrounds them.

Reflecting on our experiences, how often have we found ourselves on a path that didn't quite feel right before we mustered the courage to change our direction? While we may not always have control over our surroundings, we can always take the steps needed to adjust our path. It's about having the courage to seek the right way forward. My father's words ring true; you don't have to be a hero to change the world. Simply aim to make a difference with the best intentions.

# Chapter 2:
# Sports

Slap, Pull, Observe, Release, Tap, and Shoot

One of the foundational lessons taught in the Army is the acronym SPORTS, which stands for the steps involved in clearing a jammed weapon. Each step not only serves a technical purpose but also offers valuable life lessons.

Slap the Magazine: When faced with a jammed weapon, the first action is to slap the magazine. Life, too, will always present us with unexpected challenges or issues that can feel overwhelming. In these moments, it's crucial to maintain our composure, remain calm, and confront the situation head-on without panic.

Pull the Charging Handle Forward: This action involves pulling the charging handle to prepare the weapon for action. Similarly, every day we must gather our strength, pull ourselves together, and face the day with determination, regardless of what obstacles lie ahead.

Observe the Chamber: Taking a moment to observe the chamber of the weapon is essential for understanding the immediate situation. Likewise, in life, it's important to take a step back and assess what's in front of us, gaining clarity on our circumstances and the realities we must confront.

Release the Bolt: Releasing the bolt signifies the need to let go of the past to move forward. We all carry experiences and memories that can weigh us down, but to grow and progress, we must learn to release those burdens and embrace the future with an open heart.

Tap the Forward Assist: This step reminds us that sometimes we need a little help getting things back on track. In life, no one achieves true success entirely alone; we all benefit from the support of others friends, family, mentors and acknowledging this collaborative spirit is key to overcoming hurdles.

Shoot: Ultimately, after ensuring everything is functioning properly, it's time to shoot. This symbolizes the importance of re-engaging in life with enthusiasm and courage. Just as a soldier cannot afford to operate a weapon that is jammed, we must not allow our fears or setbacks to paralyze us. We should always strive to get back in the game, ready to face whatever comes our way.

# Chapter 3:
# Life's Hack

By embracing these principles, we can effectively navigate the challenges of military life and the complexities of our personal journeys with resilience and purpose. My mother taught me the importance of forgiveness, which is essential in life. Each of us has unique responsibilities that shape our paths, and it is in these moments of challenge that our true character emerges.

Life often unfolds in unexpected ways, embodying Murphy's Law: if something can go wrong, it likely will. Throughout my years of dedicated service to my country, I've learned firsthand that not every day is filled with the charm of a holiday. Yet, amid the toughest trials, I have stumbled upon moments of joy that shine brightly like stars in a darkened sky.

This journey has led me to reflect on the metaphorical scars and wounds we all bear. As I look back, I recognize how the structured discipline of military life has profoundly shaped me, filling my existence with purpose and resilience. Confronting the challenges of my childhood became the catalyst for my calling to serve others. Traveling across diverse landscapes and cultures has granted me a rich tapestry of experiences, allowing me to appreciate my own life from the unique perspectives of those I encountered along the way.

We don't need to experience combat to understand struggle; each of us faces our own battles at varying levels every day. I always encourage my daughters not to escalate minor disagreements into major conflicts. In a world dominated by social media and rampant gossip, it's easy to amplify small issues. People frequently share their struggles online, prompting me to consider how our lives would differ if social media had existed in the 1970s. It's in these moments that you should apply SPORTS to your life.

**"Do not feel lonely, the entire universe is inside you." -Rumi**

It took me nearly 30 years to understand that I have the power to shape how the universe affects me. Unfortunately, many people today find themselves absorbed in social media or surrounding themselves with content and individuals that offer little value to their lives.

The state of mental health in Western society can often be influenced by how we perceive ourselves based on the opinions of others. Life is about accumulating positive experiences; the more we engage in positivity, the better equipped we become to handle life's challenges. Conversely, dwelling on negativity can make the world feel overwhelming and dark. Throughout my life, I've confronted numerous challenges that have at times felt like daunting battles. However, self-motivation is crucial for overcoming those moments that seem insurmountable. We all have the potential to grow stronger and more resilient with each experience.

For years, people have been curious about the reason behind my perpetual smile. I often respond that it stems from an unwavering commitment to pay it forward to the next generation. But what does that truly mean? I like to say that I am dedicated to serving the 12.5 million souls who never had the chance to live their lives to the fullest. Each smile represents a reminder of their potential, inspiring me to carry their hopes and dreams into the future.

**"The transatlantic slave trade involved the purchase by Europeans of enslaved men, women, and children from Africa and their transportation to the Americas, where they were sold for profit. Between 1517 and 1867, about 12.5 million Africans began the Middle Passage across the Atlantic, enduring cruel treatment, disease, and paralyzing fear aboard slave ships. Of those, about 10.7 million survived, with about 40 percent of them going to work on sugarcane plantations in Brazil. Most others labored in the Caribbean, while about 3.5 percent ended up in British North America and the United States. In total, an estimated 388,000 Africans landed alive in North**

**America, and about 140,000 of these came to the Chesapeake Bay region."[1]**

Contemplating the deep significance of this citation, it's hard to envision a person who wouldn't take a moment to reflect on their own existence and the intrinsic worth of what they hold dear. I frequently remind myself of the 12.5 million remarkable individuals whose diverse stories and experiences will never be told.

I aspire to embody the best version of myself, and I hope that this sentiment resonates with you as well, encouraging you to pursue excellence and authenticity, regardless of your background or beliefs.

The resilience demonstrated by individuals navigating challenging life circumstances and environments has instilled in me the importance of refraining from undue complaints. I have learned to pursue my aspirations and objectives actively. Each individual possesses a unique role within the vast landscape of life; however, we often postpone our ambitions for trivial reasons. Consider your experiences at the age of 20 and the significant moments that may have been overlooked. It is not mere reflection that breeds regret, but rather the absence of decisive action.

I have been influenced by a generation that faced significant obstacles in establishing their place in society due to biases and discrimination. This understanding has shaped my approach to focus on factors within my control, rather than attempting to alter the behavior or beliefs of others.

There is an old saying that we pass on what we do not know as well as what we do know. More often than not, we share what is visible and overlook the vast 1% of the unknown. We play it safe in life. To reach the next level, we need vision and strong support from others. Let's be real: most of us had parents focused on retirement and pensions, missing the opportunity to build wealth for future generations.

---

[1] (Encyclopedia Virginia n.d.)

Regardless of color, race, or gender, we often overlook the invaluable lessons of our ancestors. We bear a profound responsibility to create a legacy for future generations. The 1970s mantra "live for now because you will be gone tomorrow" was a call to consume without considering our impact. I urge you to reflect on your family's history and recognize those who have left a meaningful legacy in your lineage. Too many people choose the easy path and later lament their circumstances. We must return to the essence of SPORTS in life, which means eliminating the "good-idea fairies" that distract us. The universe is full of those who dictate how to live while struggling with their own lives. The greatest fallacy is listening to others at the expense of our own vision and ideas.

# Chapter 4:
# Survivor of a Darwinian Environment

In my first assignment with the Army, I found myself under the thumb of a passive-aggressive superior a senior leader devoid of effective coaching or mentoring skills. The atmosphere was thick with tension as I navigated a chain of command that seemed to be operating in survival mode. It felt as though everyone was solely focused on their own ambitions, barely sparing a thought for the junior officers struggling to find their footing.

It quickly became apparent that I was in a Darwinian environment where only the most adaptable would thrive. I understood that the trajectory of my career hinged on my ability to navigate this toxic landscape; it could be a brief, disheartening stint or a fulfilling journey, depending on my choices and actions.

As a cadet, I was instilled with the principles of teamwork and the tenets of effective leadership. Nevertheless, it became evident that our discussions had inadequately addressed the insidious nature of toxic leadership and the self-serving individuals who often occupy such positions of authority. I would like to emphasize that my reflections do not represent a sweeping generalization; I have experienced numerous commendable assignments within the Army, during which the core values were prominently demonstrated. However, my initial entry into the Army was characterized by challenges arising from the institution's efforts to balance the second stage of social integration and the enhancement of female leadership.

I once believed that everyone held my best interests and the mission in high regard, as the military instills integrity and loyalty as core principles of service. Over time, I discovered that this truth varies with each individual. We all begin our journey in service with a sense of naivety, but I soon realized that not everyone shares the same commitment to integrity and loyalty.

As a 25-year-old soldier with less than two years of experience in the military, I quickly recognized that the military environment was not as idyllic as often portrayed. During this period, the Department of Defense was grappling with the aftermath of the Vietnam War and the Gulf War, also known as the Hundred Day War. Many military services were undergoing a significant identity crisis. Over the following two decades, the Army introduced several slogans, but none seemed to resonate effectively.

I observed firsthand the challenges facing the organization due to leadership failures at multiple levels. The Army rolled out various programs aimed at addressing fundamental issues, all within the outdated framework of the Air-Land Battle doctrine. It became clear that the Army was still dealing with a significant, unaddressed problem: the hidden legacy of segregation. Despite President Truman's efforts to desegregate the Army on July 26, 1948, I believe that deep-rooted racism and biases persisted long after my entry into military service in 1990.

As the Army adapts to the dynamics of a rapidly changing society, there are opportunities for leadership to evolve further in terms of diversity, both culturally and ideologically. It appears that some senior military leaders may benefit from broadening their perspectives to enhance the representation of women and people of color in key positions. While discussions around inclusivity are prevalent, translating these conversations into impactful actions remains a challenge. Striving for an All-American military can be a shared goal, reflecting the diverse society it serves. It's important to recognize that the dialogue surrounding merit-based selection versus diversity initiatives is nuanced and can vary greatly based on individual perspectives.

The challenges encountered by the Department of Defense were further intensified by a strategic decision made in the early 1990s to streamline military infrastructure and personnel. This initiative created a competitive atmosphere among service members. As a result, an informal phrase emerged within the ranks, "We eat our young," which reflects the challenging dynamics that arose during this period. A noteworthy observation from this time was the prevalent notion that access to knowledge was synonymous with power, highlighting the vital role of information and expertise as influential factors within the military organization.

The U.S. Army underwent a significant transformation during a period marked by intense political scrutiny and evolving military dynamics, effectively becoming a survival of the fittest within a highly-charged cultural landscape. As the institution shifted toward a micro-centric, data-driven enterprise, essential concepts such as health and welfare became mere buzzwords. These terms were often prominently displayed in PowerPoint presentations, creating an illusion of importance rather than reflecting a genuine commitment to prioritizing the well-being of service members.

In the 1990s, the military faced the complex and challenging decision to reduce the number of Captains and Majors, many of whom were on the cusp of taking on vital leadership roles within the Army. This decision, while aimed at streamlining the ranks, resulted in the removal of experienced officers a move that underscored the pressing need for the cultivation of new leadership and innovative strategies. As the landscape of military demands evolved, the decision to downsize previously promising leaders created noticeable gaps in the leadership structure. However, it simultaneously opened avenues for the Army to invest in nurturing emerging leaders who would need to navigate the complexities of future operations, including the anticipated two-front war scenario just a decade later. This transitional period can be viewed as a pivotal moment for the Army, positioning it to rethink its leadership foundation and develop strategies for ongoing and future military operations.

In a notable response to the changing dynamics, the Department of Defense (DoD) decided to extend the tenure of senior officers in high-ranking positions, resulting in a prolonged "shelf life" for these individuals. This strategy, while perhaps well-intentioned, seemed to overlook the shifting needs and realities of a modernizing military force.

As a result, this decision inadvertently facilitated the persistence of outdated practices and ideologies for an additional decade, which hampered the Army's capacity to adapt to emerging challenges and threats. The widening disparity between innovative new leadership eager to adapt and established command structures entrenched in outdated thinking contributed to stagnation in both strategic thinking and operational effectiveness during a crucial phase of military evolution.

The events of September 11, 2001, would soon challenge this status quo and prompt a reevaluation of military practices and reconstruction efforts, highlighting the critical need for modernization and a fresh approach to leadership and ideas.

**"We cannot solve our problems with the same thinking we used when we created them." -Albert Einstein**

# Chapter 5:
# War Shifts Perspective

The Department of Defense's restructuring and force reductions in the years following the Cold War were inadequately prepared for the unprecedented events of September 11, 2001. While these restructuring efforts were not directly responsible for the attacks, they served as a critical wake-up call, underscoring the urgent need to retain skilled warfighters within our ranks and invest in their future potential. During the 1990s, a significant reduction in the number of Captains and Majors led to a concerning knowledge gap within the officer corps. This shortfall not only diminished the immediate operational readiness of our forces but also highlighted the vital importance of nurturing talent, fostering leadership, and embracing diverse perspectives. The lessons learned from this period stress the necessity of maintaining a robust pipeline of experienced individuals who are equipped to navigate the complexities of modern warfare and adapt to rapidly changing global security challenges.

Today Army has a unique opportunity to transform and evolve, leveraging the lessons of the past as it navigates the challenges of the future. As society continues to advance with new technologies and cultural shifts, the military can rise to meet these challenges and redefine its place in a changing world. By fostering innovation and breaking free from the status quo, the armed forces can tap into the strength of their diverse insights and capabilities.

**"The measure of intelligence is the ability to change" -Albert Einstein**

The dawn of the two-front war on 9/11 not only altered the course of my life but also fundamentally transformed the American military landscape. After enduring a decade of strategic force reductions, the military found itself urgently seeking to replenish its ranks within just three years, drawing back those seasoned warriors who had bravely served in the First Gulf War and even in Vietnam. These veterans infused the military with a soldier's mindset, enriched by their extensive experiences and the innovative concepts they had gleaned from their time in civilian industries. They resurrected critical skill sets, particularly in conducting Low-Intensity Conflicts alongside the evolving complexities of a two-front strategy known as Air-Land Battle.

These non-political soldiers revitalized the ethos of a blue-collar work ethic long cherished in military history. The chaos and rigor of war have a profound effect on leadership, swiftly revealing the true mettle of those who lead. I observed commanders being relieved and reassigned at an astonishing pace, highlighting a stark irony: many data-driven commanders lacked the fundamental skills necessary to effectively arm, equip, and deploy troops.

I have come to refer to this tumultuous era as the rise of the new Pattons, juxtaposed against the decline of the analytical Eisenhowers. During my initial tour in the Middle East, I was privileged to serve alongside several Vietnam veterans who had lived through the harrowing Fall of Saigon. With their extensive experiences from previous tours in Vietnam, these combat veterans imparted invaluable lessons about the critical nature of timely decision-making. I will forever remember the fervor and enthusiasm of the soldiers during our intense training leading up to deployment.

However, just two months into our mission in Iraq, the stark realities of war emerged, casting a somber shadow over our initial optimism. I bore witness to profound fear and heartbreaking loss during my first tour, and I came to understand a fundamental truth: war irrevocably changes people whether for better or worse, no one emerges unscathed.

**"Man cannot possess anything as long as he fears death. But to him who does not fear it, everything belongs. If there was no suffering, man would not know his limits, would not know himself." -Leo Tolstoy, War and Peace**

# Chapter 6:
# The Ability to Act

The profound truth that war transforms individuals, whether they find themselves in the heart of a combat zone or far removed from the frontlines, is an undeniable reality. In the midst of complex and chaotic environments, the imperative to take action becomes unavoidable. This catalyst compels individuals to confront their tipping points, igniting a transformative journey of self-discovery. As they navigate the tumultuous waters of conflict both within and around them their perspectives shift, revealing insights into their own resilience, vulnerabilities, and the essence of their humanity.

Life invariably unfolds a tapestry of significant challenges, and often the greatest obstacle we face is not merely our circumstances but our own inclination to complain rather than confront the realities before us. The journey to overcome inertia begins with a single, purposeful action. I made the conscious choice to stop squandering precious time and, in that moment of clarity, unearthed a powerful motivation to seize control of my own destiny.

During my second year of college, a pivotal realization dawned upon me: I had to embrace responsibility for my situation, for I understood that no one was coming to my rescue. Absent was the safety net that many of my peers took for granted, and I found myself navigating through a landscape devoid of the kind of unwavering support that others experienced. This awakening sparked a profound transformation within me, igniting a fierce determination to chart my own path forward.

I realized that successful individuals either begin at the forefront or must fiercely pursue their peers. Starting from behind, I was ignited by an unwavering determination to reach the front. I consciously distanced myself from negative environments and stopped undermining my own progress. Instead, I centered my focus on nurturing the ambition within me that drives me toward becoming the best version of myself.

It's easy to let others influence the direction of our lives. When we start to accept negativity, it can weigh us down. It's important to find our own path in the world. There's a sense of excitement and joy that comes from creating a thoughtful plan for our future. Many struggle with the courage and self-discipline required to step outside their comfort zones, and that's perfectly understandable.

When I made the decision to raise my standards, I found that some friendships faded away. It was a difficult realization, but being surrounded by complacency pushed me to seek something more fulfilling. Time continues to move forward, often quicker than we would like, and it's natural to feel that pressure. I've shifted from being a passive observer of time to someone who actively seeks to make the most of it, knowing that every moment counts. I hope others can find the strength to do the same for themselves.

Pay close attention to those who wholeheartedly wish for your success, ensuring their unwavering encouragement surpasses any lingering doubts within you. As I reflect on my journey, I feel incredibly fortunate to have crossed paths with three remarkable individuals: Retired Major Wayne Heard, a seasoned veteran of the Special Forces; Dr. Charles Thomas, a brilliant mind dedicated to advancing knowledge; and Retired Major General Jessica Wright, a trailblazer in leadership and resilience. Each of these extraordinary figures has left an indelible mark on my life, instilling in me a profound sense of purpose and clarity of focus.

Major Heard instilled in me the critical importance of attending to the details of any mission. He emphasized the necessity of evaluating the environment thoroughly and making sound decisions, even when under significant pressure. To achieve success in either life or military endeavors, one must commit one's physical energy to the mission at hand.

Dr. Thomas imparted the most significant lesson: the importance of self-belief, even in the absence of external validation. The hours spent playing pick-up basketball and engaging with his lectures transformed my perspective. He once advised me that I am more than what I perceive, urging me to discover my true potential.

Retired Major General Wright served as an extraordinary coach and mentor, profoundly shaping my personal and professional development. Her unique ability to balance compassion with firmness created an environment where individuals felt both supported and challenged to excel. Through her thoughtful guidance and unwavering commitment, she inspired countless individuals to pursue their highest potential and strive for the best versions of themselves.

I have particularly benefited from her strategic insights and constructive feedback, which have been instrumental in my growth. To express my gratitude for her lasting impact, I make it a priority to convey my appreciation through thank-you notes. This practice not only honors her contributions but also reinforces the importance of acknowledging those who play pivotal roles in one's journey. Major General Wright's mentorship has truly become an integral part of my life's path, influencing my values and aspirations.

Throughout my career, I eagerly sought wisdom from a multitude of individuals, but a select few like the three remarkable mentors I hold in high regard profoundly influenced my early life. It's crucial to acknowledge that, although our past experiences may appear chaotic, akin to a jumble of scrambled eggs, we possess the remarkable ability to transform them into something purposeful, much like crafting a perfectly cohesive omelet. We should not perceive failure as an insurmountable barrier; instead, it serves as a vital reset a precious opportunity to recalibrate our paths. With every setback, I've gathered a treasure trove of invaluable lessons that ignite my personal growth. The real question is: do you have the grit and determination to face life's challenges boldly? All too often, we overlook our own capabilities, distracted by the tempting buffet of possibilities, rather than concentrating on what's right in front of us what is truly on our plate.

# Chapter 7:
# Vision Transforms Life

**"Peasants were worried about the future not just because they had more cause for worry, but also because they could do something about it." -Yuval Noah Harari**

I genuinely believe that individuals with a visionary mindset frequently encounter misunderstanding and skepticism from those who adhere to conventional thinking. Throughout my journey of self-improvement and personal growth, I often felt as though invisible barriers were strategically placed in my path to hinder my progress. The diverse content I encountered on social media whether inspiring success stories or discouraging narratives only reinforced my perception that some people might actively try to obstruct others' pathways to achievement.

As I concluded my military career, I was fortunate to meet an extraordinary array of individuals from various backgrounds, each imparting valuable lessons that reshaped my understanding of what truly matters in life. These interactions illuminated the reality that challenges and adversities can serve as catalysts, compelling us to seek our inner balance and discover a renewed sense of purpose. This profound realization deeply resonates with my own experiences, reminding me that growth often emerges from struggle and that resilience can be a powerful tool in surpassing obstacles.

Our environment plays a crucial role in shaping us. It raises the question: why do so many people give up on life before fully experiencing it? Life can feel daunting when we assess our surroundings and inner circles. Many find themselves in situations where government assistance becomes a way of life. However, each day is an opportunity to change your circumstances. When you endure hardships and emerge stronger, you unlock the potential for transformation. I've observed that those who say "I don't care" often face greater challenges. A life filled with misery is easily obtained, requiring very little effort. To truly thrive, we must embrace the principles of SPORTS in our lives, as this leads to a more fulfilling existence.

Throughout my experiences in both the military and civilian sectors, I have faced a range of challenges that have significantly shaped my professional journey. I've learned that visionaries often meet resistance from those who don't grasp the broader goals. From an early age, I recognized that my perspective frequently diverged from that of my peers. Despite the skepticism surrounding my potential for success, my experiences have demonstrated that such perceptions are often misguided. It's well established that our childhood experiences both triumphs and setbacks greatly influence our outlook on the future. More importantly, how we learn to navigate these experiences can profoundly affect our growth and the opportunities we create for ourselves.

To young professionals, I advise this: your vision isn't inherently flawed; rather, the audience you're addressing may lack the insight to appreciate its value. Don't let discouragement take hold many people won't believe in something until it's already achieved. You are a person of worth, and your ideas are significant, but sometimes you need to forge your own opportunities. Life will challenge you relentlessly if you don't confront it head-on. There is no need to apologize for trying to make the world a better place. The crucial lesson in life is not to wait for others to believe in you; it's vital to believe in yourself and to problem-solve in real time. Many individuals never understand why they keep failing or why they aren't growing. I simply tell them to reflect on their internal history, and they will likely discover the root of their challenges.

**I love this quote: "The best thing about the worst time in your life is you get to see the true colors of people."**

What is the next level of life? I ask you, are you ready to explore it with me? Today, I want to reach deep into your soul and illuminate the two powerful emotions that can hinder even the most visionary among us: Fear and Anger.

Fear can be paralyzing, an instinctive response that keeps us from venturing into the unknown. It whispers doubts, magnifies worries, and often stops us short of our true potential. On the other hand, Anger, while it may seem justifiable at times, can breed resentment and cloud our judgment, particularly towards those who confront their fears head-on and find success.

To navigate through life, we must learn to embrace uncertainty instead of shying away from it and to channel our anger into motivation rather than letting it turn us against others. Only then can we begin to transcend these emotions and reach new heights of personal and collective growth.

Achieving the rank of General involves navigating through various emotions, most notably determination and resilience. It is essential for individuals to embrace the concept of creating opportunities for themselves. I made it a point to volunteer for every mission that arose, recognizing that these experiences play a crucial role in shaping one's personal and professional development. This journey requires a lifelong commitment to hard work and personal growth.

What is the most valuable contribution you make to the world? If you find yourself contemplating what the world can offer you or why you should reciprocate, you are settling for mediocrity. There is one individual who wields control over your circumstances: you. Occasionally, we entrust this control to inappropriate parties. The key to a successful life is to never relinquish control to others, as they lack your vision and determination to influence the environment.

**"I am not a product of my circumstances. I am a product of my decisions." -Stephen Covey**

At the end of the day, it's you who holds the power to see the world through your unique lens, and only you can redefine the foundation of your vision. You must have the courage to bet on the person staring back at you in the mirror. Remember, when others can see that your vision aims for the greater good, you'll be able to inspire and uplift those around you. Playing it safe only keeps you stuck in the same place year after year. Every risk you take in life unlocks a new opportunity. The only thing worse than facing failure is realizing you didn't fully pursue your vision and potential. Embrace your journey with confidence.

**"Adapt what is useful, reject what is useless, and add what is specifically your own." -Bruce Lee**

# Chapter 8:
# Chasing the Mission

**"The only person you are destined to become is the person you decide to be." -Ralph Waldo Emerson**

Nobody's coming to rescue you; it's time to take charge! In the thrilling quest for success, understanding your mission is key. First, embrace your weaknesses and commit to learning how to overcome them. One of the most impactful lessons I took away from my Army experience was the military decision-making process. It's an incredible framework that transforms young minds, highlighting the exhilarating difference between playing chess and playing checkers. Too often, we find ourselves playing checkers, letting our feelings dictate our decisions and leading us down impulsive paths. However, when we pause to survey the vast landscape of our circumstances carefully, we elevate our perspective to the intricate game of chess. Here, we become strategists, contemplating each move with intention and foresight, prepared to navigate and conquer life's myriad challenges with unwavering precision and purpose.

When you commit to your life's mission, it's imperative to stay laser-focused on the elements and components that will lead you to your ultimate goal. This may seem paradoxical life doesn't truly have an endpoint but you must keep pushing forward regardless. Evolving and learning should be non-negotiable. Your self-worth is defined by the person staring back at you in the mirror, not by what others say.

If I had allowed some of my teachers and friends to dictate my path, I would have been deemed a total failure. Many of you reading this might realize that you stopped dreaming a long time ago. It's easy to point fingers and blame others for the choices you've made, but at some point, we've all abandoned parts of our life's mission. Today is your moment to stop the blame game and start evolving. Nothing worth having comes easily, and I'm the first to tell you that this journey isn't easy.

**The military decision-making process (MDMP) is an iterative planning methodology to understand the situation and mission, develop a course of action, and produce an operation plan or order (ADP 5-0).**

As we journey through life, we constantly live out our stories, yet so few of us take the time to write them down. It's enlightening to realize just how much value each person holds not only for themselves but also for the world around them. Imagine if more individuals stepped away from the endless scroll of social media and dedicated that energy to their local communities; the ripple effect could be profound, transforming not just their own lives but the lives of countless others. Your true mission in life is to build a legacy. But how can you achieve that if you're distracted by the allure of material possessions?

The key lies in nurturing your inner self, which in turn shapes the outer expression of who you are and influences those around you. Life presents us with our current situation (the reality we face) and our mission (the outcome we aspire to reach). In between, we engage in a dynamic process of planning, assessment, and adaptation. Resilience in the face of setbacks is crucial; many falter or settle when encountering obstacles. Remember, no one's journey is without challenges even those who seem to have it all together face their own trials.

It took me until my thirties to truly grasp that my life's mission was about making a positive impact on others. Yes, it sounds cliché, but the significance of this realization cannot be overstated. I believe that every good thing in life comes with a price we must pay each day. When you're genuinely invested in something, it can weigh heavily on your emotions.

To uncover your own mission, you must confront your baggage. Each of our life experiences shapes our perception of the world and our relationships, often influenced by the people who came before us. For me, processing my major baggage involved grappling with the fact that my biological father never acknowledged the mental pain he caused others.

So, let's embark on this journey of self-discovery together because each step we take toward understanding ourselves not only enriches our lives but also empowers us to uplift those around us.

I would meet my biological father years later and discover he had a troubled childhood. The turbulence of his upbringing led him to evade responsibilities and accountability. He chose the path of flight over fighting for what's right. His lack of accountability significantly influenced my perspective, drive, and character. I quickly realized that many successful people share similar backgrounds. As a father, the concept of a laissez-faire approach to parenting is difficult for me to grasp. Unfortunately, some individuals never overcome their environment and carry childhood setbacks throughout their lives. Yet, there comes a pivotal moment when tomorrow is no longer certain; only today truly matters. I've come to understand this deeply through my experiences in war and witnessing the fragility of life.

Becoming a father brings a significant awareness of the responsibility to create a meaningful legacy for future generations. This role encourages individuals to view life from a perspective that recognizes its transient nature. Time, much like a beautiful summer day, is fleeting, and once moments pass, they cannot be regained. Therefore, it is crucial to appreciate each moment and make the most of time spent with loved ones. Cherishing these interactions not only strengthens relationships but also contributes to a lasting impact on those we care about.

I understand that we all carry our own burdens. It's easy to let our struggles weigh us down, sometimes even preventing us from embracing the journey ahead. Some people find it challenging to release their baggage, which can lead to a life of unfulfilled potential and heartache. It pains me to think of that path, and I hope to inspire you to find healing and build the legacies your loved ones deserve.

This reflection takes me back to the military decision-making process, which emphasizes clarity and purpose. What is your overarching mission in life, the guiding principle that shapes your actions and choices? Have you taken the time for a thorough self-assessment to understand your strengths, weaknesses, and aspirations truly? Furthermore, consider the company you keep. Are you surrounded by individuals who inspire, challenge, and support you, adding genuine value to your journey?

**"Be fanatically positive and militantly optimistic. If something is not to your liking, change your liking." -Rick Steves**

I believe destiny will pursue anyone willing to be caught. I recall a conversation with a one-star general in Germany when I was a second lieutenant. During a formation for an exercise in Garmisch-Partenkirchen, he asked me, "Son, what are your Army goals?" With only one year in the military, I looked him in the eye and said, "I want to be an Airborne Ranger and a general like you." He smiled and replied, "Good luck with that; it's probably not going to happen." As he moved to the next junior officer, he asked the same question but received a different response. He instructed his aide to get that officer's name for his mentorship program. After the formation broke, a major approached me and said, "Do you see anyone like you with a star?" His smile sparked something in me that I cannot fully explain.

I wrestled with his comments, but in that moment, I knew I must persevere. Only those who are willing to risk the journey will ever discover their future. The moment you decide to pursue your dreams, it can feel like a lonely path. No friend or battle buddy can walk that journey with you.

**"Even though I walk through the valley of the shadow of death, I will fear no evil, for you are with me; your rod and your staff, they comfort me." -Psalm 23:4**

Upon entering the headquarters buildings, I noticed that none of the individuals depicted on the walls looked like me. This realization motivated me to provide a different perspective for others. I dedicated extensive hours to mastering my role and comprehending all three phases of the Army: tactical, operational, and strategic. I pursued knowledge with an intensity that made it my primary focus. As I deepened my understanding of military concepts, these insights began to influence my personal life as well. Ultimately, through this journey, I established myself as a subject matter expert by the time I achieved the rank of Lieutenant Colonel.

This challenge loomed large, demanding not only immense determination but also a deep personal reflection. As I navigated through the complexities of my complexion and the ingrained biases faced by officers of color, I became resolute in my commitment to ensure that these obstacles would not derail my quest for excellence. It was in the quiet corners of the base library and the bustling atmosphere of the motor pool that I began to uncover the profound significance of exceeding the expectations set by my peers. Each moment spent immersed in books or honing my skills felt like an investment in my future, even when I wasn't fully aware of its implications. To achieve my aspiration of becoming the best, I realized it was crucial to cultivate a knowledge base that not only matched but also surpassed that of my superiors.

**"The most important step a man can take. It's not the first one, is it? It's the next one. Always the next step." -Brandon Sanderson**

I dedicated a decade of my life to developing my skills and preparing for new opportunities. Embracing the unknown requires a leap of faith. One of the most important influences in my life has been the person I see in the mirror. While my journey was certainly not without its challenges, I recognized that these obstacles were steppingstones to growth. The first two years in the Army presented significant challenges but offered valuable lessons in resilience. I encountered various biases, including confirmation bias, gender bias, and microaggressions, which ultimately strengthened my resolve and determination to succeed.

During my initial assignment in the Army, I had the distinct honor of serving under a female commander, along with several senior staff members who were also women. I joined the support section, led by an accomplished female Major. As a newly appointed Second Lieutenant, my primary responsibility was to develop a more efficient strategy for revamping the unit's budget. Throughout this process, I sought guidance from my section chief on numerous occasions to refine my presentation for the Brigade Commander and to discuss key data points.

While my section chief often emphasized her rank as a Major relative to my status as a Second Lieutenant who had recently completed the Officer Basic Course, I understood that my background in a military environment particularly my stepfather's extensive career in logistics provided me with valuable insights and a unique perspective.

Recognizing the importance of networking early on, I followed my stepfather's advice, who was stationed at the nearby Air Force Base. He suggested that, in the absence of direct guidance, it would be advantageous to seek out subject matter experts. This led me to connect with an elderly gentleman in our building who often went unnoticed because of his reserved demeanor. He had been in Germany since the early 1970s and had played a crucial role in establishing the budget process for the entire base and command.

I was fortunate to receive his guidance. He helped me develop an exceptional product and advised me to rehearse my presentation until I knew it inside and out. When the day came for me to brief the commander, my section chief asked me to go first. I was aware of the informal process known as "stomp the chomp," in which senior leaders rigorously question those presenting. However, I felt prepared and confident in my knowledge, believing I had a solid grasp of the material perhaps even more than my commander and section chief.

I did not disappoint; my briefing was flawless. However, my section chief was ill-prepared, and what happened next shocked me and changed my perspective on leadership forever. Unexpectedly, I found myself called into the commander's office, where I was put at attention in front of three female officers, including the Personnel Officer, my Section Chief, and the Commander. The Commander expressed concerns about the budget and its requirements, suggesting that I had inadvertently set my section chief up for failure. After a thorough 30-minute discussion, I was finally given the opportunity to speak, having endured a rather one-sided and challenging exchange.

I had kept detailed notes in a journal that outlined the budget and areas of concern, which I knew would add context to the conversation. The Commander indicated that perhaps I wasn't suited for a career in the Army and encouraged me to consider other options. In that moment, I sensed an undercurrent of gender bias and a dark culture within the command. As I left the office, the Personnel Officer, who was African American, made a poignant remark about the importance of support among leaders, even in difficult situations.

Reflecting on this experience, I recognized the non-supportive environment and the challenges these female officers might have faced on their own path to leadership. While acknowledging their struggles, I firmly believe that gender biases have no place in an organization that has made strides towards inclusivity.

Ultimately, despite the attempt to undermine my passion for the Army, I learned that it was essential for me to focus on my growth rather than the negativity of others. Shortly thereafter, a friend from another command reached out and recommended me for a position that suited my skills. After some initial resistance from my former unit, I successfully made the transition.

In my new role, I thrived at all levels and embraced every opportunity that came my way. This experience taught me the value of resilience and the importance of seeking environments where I could truly excel. This is why my life mission involves exploring different roads and paths.

# Chapter 9:
# The Shift

**"Take the first step in faith. You don't have to see the whole staircase, just take the first step." - Martin Luther King Jr.**

It wasn't until I read General Colin Powell's book that my journey took a transformative turn. I recognized that being an African American Officer in the United States Army presents unique challenges. Recently, the current Secretary of Defense, General Lloyd Austin, shared his experiences with the micro biases he encountered as a Lieutenant Colonel. Unfortunately, many officers of color continue to face these challenges in various forms today. The remnants of cultural and societal stereotypes persist, echoing the past in ways we must acknowledge.

Understanding the history and struggles of my grandparents has provided me with a powerful advantage. It instills in me a no-excuse mindset, empowering me to pursue success without attributing my failures solely to race. My intent is not to offend, but to be clear: the African American officer in the 21st Century does not confront the same obstacles as our predecessors. There are no longer open attacks on character without means for rebuttal. Today, opportunities are available for all who seize them, yet the shadows of old challenges remain.

The military's hierarchical and rigid structure has its complexities, often lacking a clear, open-door policy. As I often remarked, "We have plenty of policies and guidance, but there are so many layers." I had to break free from being a mental victim of society and find my place at a predominantly monochromatic table. I vividly recall an officer questioning my presence at a meeting for general officers shortly after my promotion, a comment that, despite its jest, carried an underlying tone.

Throughout my thirty-year career, I faced similar situations countless times, yet I remained undeterred, motivated by those who paved the way with fewer opportunities. We have the power to change the world, one day at a time. I remember a commander, COL D.V. Childs from Macon,

Georgia, who imparted invaluable wisdom: "Racism doesn't know hard work because everyone has to respect a hard worker." Commander Childs helped me see the world through a different lens, and he remains one of the best commanders I ever had the privilege to serve under in my three decades of service.

**"In three words I can sum up everything I've learned about life: it goes on." -Robert Frost**

If you were to write down all the people who failed you and all the people who helped you, which list would be longer? One day, I began a list of those who failed me, focusing on their lack of support during my most challenging moments. Through it all, my mother, sisters, and brother stood as my unwavering foundation. In contrast, the list of those who helped me selflessly was strikingly small.

Once I let go of the frauds in my life, I began to grow and achieve success. The key to freedom lies in knowledge. History shows us that enslaved African Americans were kept illiterate out of fear that knowledge could empower them to break free from societal constraints. While there are countless ways to transform our lives, mastering knowledge is the first step. An abundance of knowledge equips a person to change the world around them. I firmly believe that I hold the power to shape my future and the futures of others.

Many remain stuck in the mindset of who belongs at the top. We should strive to become the best versions of ourselves. Life is a mission, and easy solutions are rare unless one is born wealthy. If you don't actively shape your life, society will do it for you often introducing challenges that can take decades to overcome. Life doesn't give you what you want; it reflects what you invest in it.

This is where we discover our purpose. Many fear cutting negative people from their lives; overcoming that fear leads to freedom. I quickly realized I had to stop allowing hateful individuals to control my journey. Always devise an exit strategy from toxic relationships and have the courage to follow through. I had a self-serving commander, and navigating individuals like that is sometimes necessary for success.

What complicates life is often being surrounded by individuals who settle for average. I vividly recall a pivotal moment during one of my missions when my commander assigned me a critical task, urging swift action. Meanwhile, some of my peers suggested that I take my time to plan every detail meticulously. Their voices echoed in my mind, but my intuition, honed through experience, urged me to act without delay. Trusting my instincts, I dove into the planning phase right away, mapping out strategies and contingencies that could adapt to any unforeseen circumstances. When the mission concluded, my commander expressed his disappointment with those who hesitated, firmly reprimanding them for their lack of urgency. My proactive approach not only led to a successful outcome but also served as a stark reminder of the importance of decisive action in the face of mediocrity. One officer even remarked, "You made all your peers look bad." Success rarely favors those who approach life with mediocrity. What will your tombstone read? I want mine to say, "He never wasted the time he was given."

The military's strict hierarchy can be complex, often with layers that obscure open dialogue. I had to free myself from the mental victimhood society imposed and find my place at a predominantly monochromatic table. I vividly remember an officer questioning my presence at a general officer meeting shortly after my promotion. Though it was said in jest, the underlying tone was clear.

Regardless of your circumstances, you must steer your life in the right direction. You are the General of your own life. Never relinquish control to situations that do not prioritize your best interests.

**"It is not the critic who matters, nor the person who points out how the strong man stumbles or where others could have done better. The credit goes to the person in the arena, whose face is dirty from dust, sweat, and blood. This person strives bravely, makes mistakes, and often falls short because there is no effort without error or failure. They understand great passions and dedicate themselves to a worthy cause. In the end, they know the joy of high achievement. If they fail, they at least do so while taking risks, ensuring they do not belong to the cold and timid souls who know neither victory nor defeat." - Theodore Roosevelt**

# Chapter 10:
# Nothing Survives First Contact

No matter how meticulously we plan our lives, it's a universal truth that no strategy survives the moment it encounters reality. Despite our best efforts to navigate the unpredictable landscape of existence much like trying to maneuver a joystick in a video game life often finds ways to humble us at the most inconvenient times. The phrase "no plan survives first contact" emphasizes the importance of anticipating and being flexible when facing challenges. In complex situations where obstacles arise, it becomes crucial to find clarity amidst confusion. In combat scenarios, individuals often face a pivotal choice: either fall into despair or continue to persevere. Both options can yield results.

For many of us, life can be compared to a trip to a shooting range. The targets that lie within 50 meters are relatively straightforward to hit; we can adjust our aim, refine our technique, and rely on our training. However, it's the targets positioned at 100 meters or further representing our loftiest goals and aspirations that present a far more daunting challenge.

Even with extensive preparation and practice, these distant targets become increasingly difficult to strike as life introduces dynamic variables that shift the very landscape around us. Factors like unexpected opportunities, personal crises, and changes in our environment can throw off our aim, forcing us to adapt and recalibrate in ways we never anticipated. Ultimately, this reality teaches us resilience and the importance of flexibility in the face of life's inevitable uncertainties.

Like practicing at a shooting range, I fully understand that achieving genuine success demands an unwavering commitment and relentless perseverance. I kept my eyes fixed on the horizon of my professional future, where I aspire not only to reach my goals but to elevate my ambitions beyond the limitations imposed by others. The challenges I faced as a young man were not just obstacles; they were critical elements that would refine me into a resilient warrior, capable of facing life's uncertainties.

Every day presents a fresh opportunity, much like a new session at the range, where each shot helps hone my skills. I recognize that even if my meticulously crafted plans don't survive their first encounter with reality, each setback is a chance to cultivate inner strength and wisdom. Through adversity, I can shape my character in profound and meaningful ways, emerging stronger and more prepared for whatever lies ahead. This journey is not merely about achieving success; it's about the transformation that comes from confronting and overcoming the difficulties we face along the way.

Each day, I am committed to deeply evaluating my life, reflecting on my achievements and challenges. I assess my current situation career aspirations, personal relationships, and emotional well-being to gain clarity on my journey. By recognizing my strengths and areas for growth, I can create an internal roadmap that guides me toward success. With clear goals and actionable steps, I am ready to embrace the opportunities ahead. Success is not a distant dream; it is within my grasp, and I am prepared to seize it with confidence and purpose!

My mission in life became a singular focus: to do more for others and allow God to send blessings my way if I deserve them. I never prayed for wealth, but for humility. Many people struggle with dissatisfaction in their lives because they are self-serving or live by a quid pro quo mindset. During my time in the Army, I cherished working with the infantry, who embody a powerful saying: when chaos erupts around them, their only option is to hold the line at all costs, knowing that breaking ranks could cost a life. As humans, we must also hold the line when our life's mission faces its first challenges. It's a profound truth that no mission survives first contact.

Those who embrace adaptability uncover new opportunities, while those who resist change risk stagnation in a joyless world. The result? Mission failure. So, how do we rise above this failure? It starts with crafting a strategy that aligns your current reality with your aspirations. Many shy away from self-assessment, finding it easier to blame others for their circumstances. Embrace the journey of self-discovery and watch your mission transform!

**"The leader is one who, out of the clutter, brings simplicity... out of discord, harmony... and out of difficulty, opportunity." - Albert Einstein**

I challenge you to dive headfirst into the adventure that is your life! When you redefine the rules you live by, you unlock the first step to taking control of your destiny. Don't sit around waiting for someone else to tell you it's time to move because guess what? The world won't pause for you to muster the courage to save yourself! The moment you shift your mission, you open the door to transforming your entire future. If you want to build unshakable resilience, it begins the instant you change your routine.

Life is a thrilling challenge, full of jaw-dropping moments that require us to face what we fear most. The concept of "mission" revolves around understanding your circumstances and conducting an honest assessment. Life becomes exhilarating when you choose to flip your mindset. Remember, you can't guide negative people to greatness, so sometimes you might have to blaze your own trail. Some may feel stuck in limbo, but that's not where you want to be! Embrace life, traverse through the heartaches, losses, and challenges. That's what it truly means to be alive! Get ready to reinvent your narrative and ignite the fire within!

**"Optimism is a huge asset. We can always use more of it. But optimism isn't a belief that things will automatically get better; it's a conviction that we can make things better." - Melinda French Gates**

# Chapter 11:
# Understanding Strategy

In our journey through life, we encounter a variety of battles, each testing our mental and physical resilience. The reality is, the sun doesn't shine every day. There will be moments when you feel completely isolated, and it's during these times that you must strategize. What will propel you off the sofa and away from the endless scroll of social media? What steps will you take to construct a blueprint for your future?

**"If you know the enemy and know yourself, you need not fear the result of a hundred battles. If you know yourself but not the enemy, for every victory gained you will also suffer a defeat." - Sun Tzu**

Strategy is more than just a vague notion; it's a comprehensive plan designed to achieve long-term goals amid uncertainty. Yet, many people live life defensively, lacking a clear strategy, play it safe, and miss out on the opportunities that offense can bring. From the moment we enter this world, we are surrounded by unpredictability. We have little control over the circumstances of our birth or the environments we grow up in, yet we must adapt and forge our own path through the maze of life.

There will undoubtedly be times when you feel like the world is against you. Just remember, while it's easy to slip into hopelessness, you have the power to control your mindset and emotional reactions to life's challenges. Don't let someone else's dream dictate your life unless it genuinely enriches your own. Take a cue from the military, where recruits learn to solve problems under pressure. A soldier's weapon is not only a tool but a vital lifeline always maintained and ready to protect amidst chaos.

In life, having a strategy is crucial. Those who mentor and guide others towards strategic thinking often become high achievers, leaving many to work for them simply because they lack direction. Countless people possess skills and knowledge, yet without a strategy, they remain stagnant. Today, it seems that if something doesn't yield instant results, it's deemed a waste of time. However, real success often requires slowing down and crafting a five-to-ten-year plan a long-term vision that sustains you through life's tests.

Strategy can be viewed as a dynamic discipline that evolves with our learning experiences. When we consider the creativity of a child marked by vibrant and unrestrained imagination it starkly contrasts with that of adults. This shift raises an important concern regarding our education system, which frequently limits creative potential. In many Western societies, there is a heavy reliance on standardized testing that often fails to reflect individuals' true abilities accurately. This reliance may hinder the development of creative skills crucial for innovation and problem-solving.

During my tenure in the Army, I observed how leadership can sometimes inhibit creativity within teams. My own success was largely attributed to creating an environment where non-commissioned officers (NCOs) and junior officers could express their creativity and innovate. As a leader, it is essential to foster a team atmosphere that promotes growth and development. When team members feel invested in a shared vision, they are better equipped to address external challenges with purpose and determination. Additionally, having a solid strategy is crucial, as even the best intentions can fall short without it. The ability to focus and maintain clarity is essential for success, as the world tends to reward these qualities over chaos. It's a pivotal choice: to adapt, embrace the learning process, and flourish, or to let the difficulties define us. Each experience can be a stepping stone towards a brighter future.

There was a transformative moment in my journey when I made the difficult choice to hit the pause button on my business endeavors, redirecting all my energy and passion toward completing my book. The frustration I experienced was palpable, yet I understood the profound necessity of sacrifice. I refuse to dwell on the shadow of the person I might have become; instead, I am fiercely committed to embodying the man I aspire to be. Too many individuals wander aimlessly through life, lacking a clear vision and strategy. True winning isn't merely about accumulating accolades; it's about proving to yourself that you can rise above the challenges that life throws your way.

Before I embarked on this writing journey, I faced a barrage of setbacks, largely due to a haphazard approach and a lack of planning. However, once I forged a strategic path, the words began to pour forth with a flow that felt almost effortless. I had stared down the barrel of many failures, largely due to my own procrastination, and it was in this crucible that I learned the essential truth: prioritizing oneself is the key to liberating one's potential.

This pivotal turning point prompted me to recalibrate my focus, slow the pace of my podcast, and channel my energy into the written word. It serves as a powerful reminder that, with the right mindset and a well-crafted strategy, you can navigate life's complexities and emerge stronger and more resilient than before.

**"Strategy without tactics is the slowest route to victory. Tactics without strategy is the noise before defeat." - Sun Tzu**

In the 1920s, inventors often spent a decade or more meticulously gathering the necessary components to bring their innovative ideas to life. Even after years of dedication, they typically had only one to three opportunities to succeed with their creations. Fast forward to the 21st century, where the fast-paced nature of life, influenced heavily by social media and celebrity culture, has led many to expect instantaneous results.

It's a stark contrast to the historical norm; today, the majority of individuals display a marked impatience for success, seeking shortcuts and rapid achievements. The reality is that sustainable success requires a significant investment of time and effort a principle succinctly captured in the phrase "bottom line up front" (BLUF).

In the 20th century, we propagated another misleading narrative with profound effects on our society's trajectory: the idea that the only path to success was through a traditional college education. Beginning in the early 1990s, this message overshadowed the vast array of alternative career options, particularly skilled trades, which were systematically de-emphasized in our educational systems.

This exclusion effectively narrowed the pathways individuals could use to chart their lives and build promising careers. Nowadays, many remain blissfully unaware of a crucial trio Ends, Ways, and Means that plays a vital role in constructing a strategic approach to life. Regrettably, the prevailing outcome is that countless individuals accumulate significant debt in the pursuit of an education, only to find a mismatched job market awaiting them.

The truth is stark: there are insufficient well-paying remote jobs for degree holders, while the demand for skilled workers in our modern industrial landscape remains unmet. Moreover, many graduates leave college with an illusory roadmap, often envisioning a six-figure salary as the ultimate goal. However, the real world often delivers a harsh awakening for those who lack preparation. The grim reality sets in as they realize their well-constructed Means do not lead to fulfilling Ends.

One essential strategy taught by the military yet seldom acknowledged in civilian life is the art of improvisation and adaptability. This skill set proves invaluable when the original plan falters or becomes outdated. A historical example can be drawn from the experiences of German soldiers during World War II. While most European military units adhered to a traditional top-down command structure, where soldiers maintained their positions until their officers were incapacitated, American forces embraced a different approach. In the event that a commanding officer fell, they were trained to empower the next high-ranking soldier, fostering an environment of problem-solving and initiative even amidst chaos.

The ability to improvise under pressure is a skill that has waned in today's society. Many individuals grapple with mental health challenges arising from stressors akin to those faced by soldiers in wartime, yet they often lack the coping mechanisms necessary to navigate modern adversities. In a similar vein, pursuing ambitions entails inherent risks and challenges. It takes considerable courage to advance with an untested plan and to pursue dreams that may not have a clear trajectory. This journey demands resilience and adaptability at every stage.

Strategy emerges from ideas and is shaped by human disciplines. Regardless of the context, all effective strategies require one essential element: the opportunity for evolution. It's also important to recognize that while war can be hellish, achieving peace is a continuous endeavor that requires daily attention and effort.

# Chapter 12:
# Recalibrating Life

After transitioning from military service, I founded a company with minimal financial resources. Each day requires a high level of creativity in finding innovative solutions to achieve my goals. It's often said that challenging times foster exceptional individuals, but I believe that difficult circumstances are instrumental in shaping creative thinkers. Understanding one's environment is crucial in this process. Many individuals are influenced by their past experiences and need to re-evaluate and adjust their mindset to envision a successful future.

So how do we build a plan? Some people write it down, while others say just start, and some believe in having faith that it will happen. The truth is, you must believe in what you cannot see. Imagine walking down a dark street with no lights; your mind races, sending messages about the unknown. It negotiates your possible actions. But when you begin to negotiate your life's plan, you risk heading toward a failed outcome. Ultimately, it's about taking personal responsibility and mustering the courage to take that first step.

To achieve your dreams and goals, it's essential to understand your current environment. Many individuals may not fully grasp their organization's structure, which can limit their potential for growth. Instead of merely showing up for a paycheck, take the initiative to research and learn about the pathways to advancement within your workplace. I discovered that my understanding of the Army's political structure was pivotal in crafting my strategy for success. When you commit to achieving your goals, regardless of the obstacles you face, you can create a personalized plan tailored to your journey. Success is not one-size-fits-all. Consider how much progress you could make if you embraced challenges and put in the effort to grow.

**"We all have two lives. The second one starts when we realize we only have one." - Confucius**

Many people often make impulsive emotional decisions that result in long-term negative consequences, simply because they lack a clear strategy or vision to tackle life's obstacles. In contrast, successful individuals prioritize long-term outcomes over fleeting victories. Imagine the powerful transformation you could achieve by changing your environment or creating a concrete plan for your future. For some, getting into shape can be a catalyst for a complete life overhaul. Embrace the potential for change and take charge of your life!

The truth is that the person in the mirror is often lazy and avoids extra work because they lack a solid plan. Recognizing your value isn't arrogance; it's essential. The biggest mistake is living off past successes instead of growing. The Army taught me that shortcuts can lead to failure or even danger.

In reality, many people actively choose to dance around discomfort because they find a twisted sense of comfort in drama and chaos. You must sometimes detach yourself from your surroundings and prioritize your own growth. This is where a well-defined strategy becomes essential. The moment you commit to living your second life is the moment you take control of your future. Consider this: do you know anyone who has never left their hometown? How many people do you know without a passport? Life is far too short to navigate without a clear strategy for living. Our environment shapes us, and it profoundly influences our behavior and outlook. Countless successful individuals attribute their achievements to a pivotal shift in their environment. Take charge of your circumstances and watch how it transforms your life.

During my time living abroad, particularly in the vibrant cultures of Europe, I discovered an even greater sense of fulfillment and success than I had experienced while confined to my familiar surroundings in Jack's Quarters in Bossier City, Louisiana. This international journey was not just a geographical shift; it significantly broadened my mental perspective. Immersing myself in diverse societies and viewpoints helped me cultivate a more strategic and long-term approach to my life decisions.

What stands out to me is the number of individuals who choose to remain in environments with minimal accountability. These settings might seem comfortable at first, but they can become toxic traps that stunt personal growth. It's crucial to understand that wherever you find yourself in life, you have the power to change your circumstances. Yet, with that power comes the responsibility to hold yourself accountable for your choices.

Transforming your life isn't an overnight process: it requires dedication and time to undo the patterns formed by years of poor decisions. However, if you genuinely desire a change, it's essential to critically reassess your strategy. This means setting specific, measurable goals and being willing to adapt your approach based on what you learn along the way. By doing so, you give yourself the best chance to navigate toward a more successful future.

**"One can have no smaller or greater mastery than mastery of oneself." -Leonardo da Vinci**

What does strategic planning look like in the 21st Century? It's a dynamic blend of traditional journaling and goal-setting that reflects your vision. Writing down your goals holds immense power, especially when revisited weekly. Challenge yourself at least four days a week and put yourself to the test. The more you push your limits, the higher your standards will soar. Don't settle for writing the final chapter of your life; start building a lasting legacy. The reality is that many will give up before they even begin.

To initiate your writing journey effectively, start with a small notebook, such as a Moleskine, and choose a quality fountain pen. A smooth writing instrument can enhance the experience of capturing your thoughts on paper. While there are various outlining strategies available, a useful approach is to clearly define the reasons you are pursuing your goals.

**For instance, consider the following points:**

**A. Establish a life goal, such as writing a book.**

**B. Identify challenges, such as the difficulty of getting started.**

**C. Recognize the importance of prioritizing writing time over less valuable activities, such as watching television.**

**D. Set a timeline for your project, like scheduling the completion of your first chapter and finalizing your last chapter.**

Implementing a structured note-taking process and adhering to a strict timeline can significantly enhance accountability in achieving your writing goals. This means dedicating specific hours of your day exclusively to writing, which helps combat procrastination. It's crucial to recognize that achieving challenging tasks often requires perseverance, even on days when motivation wanes. Ultimately, to transform your life, it is necessary to regularly evaluate and reassess your goals, regardless of what you're aiming to accomplish.

# Chapter 13:
# Moving On

Many people transition from this world without achieving their potential because they lack a strategic, viable plan. Countless self-improvement books exist, but they only work when applied. It's essential to confront your inner struggles before you can truly succeed. You don't need to forgive everyone who has wronged you, nor should you bear the weight of others' mistakes. I've discovered that while you don't need everyone in life, having a few good people is crucial. Once you dig deeper, you may find that your inner circle is either small or nonexistent. It's disheartening to see individuals live without purpose, hoping for a golden parachute in later years, yet never taking responsibility for their lives. The reality is, poor planning will inevitably lead to burdens we must carry.

Many people in retirement homes reflect on their lives and wish they had crafted a clearer plan. While improvising can seem like a life strategy, it often leads to unintended outcomes. In our youth, we rarely consider aging, but the reality is that with time, we come to realize how valuable a thoughtful plan can be.

The mottos of the 1970s "Make love, not war. Don't let the man keep you down. Live and let die. Hell no, we won't go." celebrate living in the moment. Yet, when those moments pass and the consequences of inadequate planning manifest, we truly understand the importance of a meaningful life journey.

The significance of our lives is often measured not by our early years but by the actions we take in the middle and later stages. This highlights the importance of having a well-thought-out plan. In the 1970s and 80s, a prevalent mindset among Baby Boomers was encapsulated in the phrase, "I am going to get mine, and to hell with everyone else." This attitude permeated pop culture and influenced many professionals during that time, reflecting the societal conflicts of those decades. In today's world, similar challenges persist regarding effective planning for the future.

Historical patterns indicate that many individuals may not adequately prepare for what lies ahead, even with readily available information. Life is inherently complex, filled with obstacles and challenges, and it is wise to navigate these difficult terrains with care.

The experiences gained from failures and setbacks provide crucial insights that can be leveraged to develop a successful strategy moving forward. A simple exercise can help bring clarity to your current situation:

1. **Spend one minute writing down your current profession.**

2. **Allocate another minute to reflect on your level of happiness in this role.**

3. **Use two minutes to assess whether your education aligns with your current income.**

4. **Finally, dedicate three minutes to outline actionable steps you can take moving forward.**

Given that you have spent five to ten years in your current situation, investing just seven minutes to evaluate and strategize can be a small yet impactful step towards meaningful change.

**"Confidence doesn't come from having all the right answers, it comes from being ready for all the questions." -Christina Nepstad**

Disclaimer: My initial vision for life took an unexpected turn. I began my college journey with the fervent ambition of becoming a war correspondent, driven by a desire to traverse the globe and delve into the complexities of international conflicts. Yet, what I unearthed was a different, profound journey that allowed me to view the world through a distinctive lens a perspective enriched by unexpected encounters and experiences.

It's truly fascinating how life intricately molds our aspirations in surprising ways. As we pursue personal growth, we navigate a series of transitions that shape our paths and redefine our goals. I want to emphasize that, through these shifts, I have remained resolute in my commitment to pursuing my dreams. My ultimate aspiration is to cultivate the best version of myself.

Success, I have come to understand, is not solely defined by moments of jubilant triumph; it also encompasses the humbling setbacks that often punctuate the journey. Each ecstatic victory, whether it be landing a dream job or completing a challenging project, fills me with pride and reinforces my belief in my abilities. Meanwhile, the setbacks like a failed attempt at a personal goal or constructive criticism from a mentor serve as invaluable lessons, forcing me to reevaluate my strategies and goals. These experiences, both joyous and difficult, are essential components of my learning process. They deepen my understanding of resilience, adaptability, and perseverance, guiding me along the often winding path of lifelong growth and exploration.

Who do you need to transform into to infuse your life with meaning? During my time as a captain, I reached a pivotal moment where I could no longer hide from my failures or shy away from those who harbored resentment towards me. I discovered the power of my inner drive, shedding the heavy cloak of fear that had held me captive for far too long. Unlike one of my peers, who continued to rely on the whims of fortune, I came to a profound realization: life is neither fair nor equal, and it's time to accept that truth. Take a moment to reflect on the countless times you've allowed the negative influence of others or your own anxieties to block your path in life. How many opportunities have slipped through your fingers as a result?

I want to emphasize that this is your moment seize it and put in the hard work now! There was a time when I pondered my own purpose in life. After watching the timeless classic, "It's a Wonderful Life" (1946), everything changed for me. In the film, Jimmy Stewart's character wishes he had never existed, only to discover, through divine intervention, the profound impact he had on those around him. He realizes that his life, with all its ups and downs, contributed to the greater good. This story powerfully illustrates that each of us has a unique purpose. It's not the length of our time here that matters most: it's the meaningful actions we take that leave a lasting legacy. Take this opportunity to make your mark!

Reflect on your perception of the world and the unique value you bring to it. The distinction between strategy and execution often lies in the ability to turn plans into reality within a predetermined timeframe. Achieving success means mastering the ability to control the sequence of events in a way that aligns with your goals. Life's proper order involves first defining a clear vision, then meticulously crafting a detailed plan, followed by a thorough analysis of your strategic decisions, and finally, executing that plan effectively.

During my tenure as a general officer, I had the incredible opportunity to collaborate with some of the most brilliant minds globally. These individuals brought a wealth of knowledge and innovative ideas to the table, inspiring creativity and problem-solving. However, I also encountered individuals who were incredibly self-centered and focused solely on their own advancement rather than the collective mission. This dichotomy reinforced my understanding that every person possesses the capacity to choose how they conduct themselves in different contexts.

Regrettably, the Army often experiences a pervasive issue known as confirmation bias. This social dynamic can be compared to the behavior of teenagers navigating social media; if one person decides to sever ties, it often creates a ripple effect, where others feel compelled to follow suit. This unspoken code of silence among peers can stifle open communication and hinder professional growth, ultimately undermining the careers of dedicated individuals who might otherwise thrive in a supportive environment. It's essential to recognize and address these challenges to foster a culture that encourages genuine collaboration and mutual respect.

I have come to appreciate the complexities of this environment in part through the insights of General Colin Powell. Throughout his career, he faced challenges, including being unjustly designated as a scapegoat for the shortcomings of others on two occasions. A particularly notable moment that resonated with me was when he received a below-average rating from a senior officer. His remarkable journey to achieving the rank of general despite that setback serves as a powerful reminder that with self-belief, one can overcome obstacles and achieve great things.

**"In real life, strategy is actually very straightforward. You pick a general direction and implement like hell." -Jack Welch, former CEO of General Electric**

Every individual carries a unique value within them, yet countless people drift through life without ever realizing or acting upon it. You were not placed on this earth merely to exist in the background; rather, you have a vital role to play in making a positive impact on the lives of those around you. If you find yourself lacking in that impact, it may stem from the distractions that cloud your vision.

To leave a mark on the world, you must first embark on a journey of self-discovery, understanding who you truly are. Life demands a well-thought-out strategy if you aspire to become a force of change. Establish guiding principles for yourself commit to hard work, cultivate gratitude, and foster a spirit that celebrates the success of others instead of harboring resentment. Be willing to loosen your grip on the aspects of your life that no longer serve you. This is a personal voyage, unique to each individual; no one else can walk in your shoes.

=Seek out what success truly means for you. When disappointment weighs upon you, recognize that a staggering 90% of that burden is rooted in choices you've made along the way. The key to navigating life's complexities lies in self-awareness and in evolving into the person you aspire to be. Life often resembles a battlefield, where challenges arise and obstacles loom large, creating an atmosphere akin to a never-ending war. To emerge victorious in these struggles, you need a strategy that aligns harmoniously with your goals and vision.

=Life presents you with a series of choices, most of which are strikingly simple: will you choose to walk the path laid before you or turn away? While there are numerous decision-making frameworks available to guide you towards success, the heart of each model centers on one essential element: you.

=These strategies flourish only when brought to life through action. Consider life as a process of sculpting a strong physique; it necessitates the right balance of nutrition, exercise, and restorative rest. In the beginning, this journey may seem effortless, but as time progresses, it demands sacrifices dedicating precious hours to workouts, swapping comfort food for nourishing choices, and prioritizing rest over fleeting entertainment.

Your life plan is as distinctive as a tailor-made suit, perfectly designed to fit your unique shape and style. When you commit to this journey of growth, you will unveil new avenues and possibilities. As long as you pledge to invest the effort required, this path of self-improvement will lead you to the success you seek, transforming your life into a masterpiece.

# Chapter 14:
# Understanding Destination

**"Death is the destination we all share; no one has ever escaped it. And that is as it should be because death is very likely the single best invention of life." -Steve Jobs**

It's essential for us to acknowledge that the moments we experience are precious and fleeting. Each day presents us with the opportunity to reflect on our lives and the reality of our mortality. Often, it is the passage of time or significant challenges, such as health crises, that inspire us to engage in deeper self-reflection.

I have been inspired by the memory of a dear friend who lost his life in Iraq nearly 20 years ago. My adoptive father, Alton G. Buggs, Senior Master Sergeant in the U.S. Air Force, often reminded me that war and death are closely intertwined. This perspective resonates with the insights shared by the late Steve Jobs, who spoke passionately about life's importance in one of his final university speeches.

As we navigate our journeys, we should consider the legacy we wish to leave behind. Each moment is an opportunity for growth and making meaningful connections. By embracing this awareness, we can inspire ourselves and others to lead fulfilling lives. What legacy do you want to create, and how can you make the most of your journey?

Life is incredibly fleeting, and it's important to cherish our time rather than dwell on what we can't change. Steve Jobs had a profound point when he reminded us of our mortality. After 30 years of experience, I've come to appreciate the importance of celebrating our achievements rather than fixating on our setbacks. The journey towards our goals can be challenging, and it's disheartening to see how many people allow excuses to hold them back from even beginning that journey.

Many individuals find themselves in a cycle of disappointment, often feeling trapped by their circumstances. A few years ago, I lost a dear friend to cancer, and I still hold onto her final words to me: "Keep moving forward." It's a poignant reminder that we truly only have the present moment. I understand that for some, the desire for success might be overshadowed by the fear of required change.

A wise colonel once shared that opportunities often come just once in a lifetime, but they usually demand that we take risks and reinvent ourselves   especially if we remain in the Army for an extended period. Life is not just about existing. Life is about moving forward with purpose and energy. Let's inspire and support each other on this journey of growth and transformation.

**Martin Luther King Jr. said, "If you can't fly, then run, if you can't run, then walk, if you can't walk, then crawl, but whatever you do, you have to keep moving forward."**

We all face moments when we must persevere and be strong-minded in the face of change. Life is not about what happens to you but how you react in times of turmoil and confusion. People live their entire lives going through the motions. Just remember, we all face moments that test our perseverance and the strength of our minds in the face of change. Life is not just about what happens to us; it's about how we respond in times of turmoil and confusion. Many people drift through life, but remember, the only steadfast companion you will have until the end is yourself.

In 2003, during the intense thrust of Operation Thunder Run, the U.S. military surged towards Baghdad, marking a pivotal moment in the Iraq conflict. Yet, amidst the chaos and urgency, several units found themselves hopelessly disoriented, victims of ineffective leadership and an insufficient grasp of the unfamiliar terrain. This predicament echoed the experiences of my unit during our deployment to Iraq in 2004. By that time, Iraq had evolved into a fortified stronghold for insurgents, transforming the once-familiar main supply routes, Tampa and Bronze, into treacherous passages fraught with danger for our service members.

I can still picture the moment my unit geared up for our deployment to Iraq. As I pored over the map, my eyes scanned the expansive arid landscapes, where the rolling earth stretched endlessly beneath a blazing sun, broken only by the faint outlines of distant hills. The only significant path marked was the main supply route, known as Tampa, a lifeline cutting through the barren terrain. It struck me how life without recognizable landmarks was akin to wandering through a vast desert, each mile blending into the next, as if I were a lost soul yearning for celestial guidance to steer me back home.

**"I can't change the direction of the wind, but I can adjust my sails to always reach my destination." - Jimmy Dean**

War is a harrowing reality, and the staggering loss of 4,492 brave service members in Iraq serves as a stark reminder of its horrors. Many of these heroes fell victim to IEDs (Improvised Explosive Devices) on two crucial supply routes. I vividly recall the night my unit touched down in Kuwait. As we stepped off the plane, it hit me that we were about to enter hostile territory. Our unit was led by an experienced Sergeant Major, SGM Corson, who had fought valiantly in Vietnam as a teenager during the mid-1970s. His powerful words still ring true: "The enemy gets a vote, and so do the elements; either can take your life." These truths underscore the unpredictability of war and the grave risks our service members face.

For the very first time, I was struck by the raw fear, palpable concern, and undeniable bravery of the men and women by my side as we prepared to deploy. I pored over the map of the Middle East, my heart racing as I absorbed reports from units already navigating the terrain. Each line on that map and every detail in those reports painted a vivid picture of a world steeped in history and conflict. The gravity of what lay ahead hit me hard dying in a foreign land was not just a distant thought; it became a stark reality.

As I reflected on the gripping tales shared by my history teachers and my father's friends legends of World War II, the Korean War, the Vietnam War, and the Desert Shield/Storm veterans I felt a surge of determination. The moment I touched down in Kuwait, everything shifted. I learned to seize the day with both hands, knowing that for so many, tomorrow was anything but guaranteed.

This awakening ignited a fire within me a realization that life is too precious to be wasted. It's time to carve out your unique path, to break free from the monotony of everyday existence. Embrace the thrill of adventure and refuse to live in circles; take charge and eliminate that Groundhog Day routine! Your journey awaits, and the world is calling!

There is a critical realization for deploying service members: all those years of training have significant value. The jokes about the long hours spent mastering land navigation or the complaints about using a manual compass fade away in the face of reality. You are now in the moment, confronting a real enemy, with the responsibility of protecting the lives of your fellow soldiers.

There are two fundamental approaches to life: Have I done everything possible to navigate what is in front of me, or did I take a shortcut? This brings me back to the concept of IEDs. Regardless of who you are, you will encounter your fair share of IEDs in everyday life. Unlike landmines, IEDs are triggered by motion devices or humans. Sometimes, the only way to avoid an IED is to forge a new path or wait for a specialized team to clear the obstacle. Yet, many people hesitate, waiting for someone else to remove complex challenges from their lives, lacking the emotional strength to confront them on their own.

War teaches us invaluable lessons, often harsh and unforgiving. It reminds individuals to pay attention to training because the consequences can be dire. The essence of life lies in discovering more about ourselves. Life-threatening situations can strip away confusion and reveal clarity. While we cannot control time, we can determine our destination after surviving trials. Failures and hardships are part of life's journey.

Before the invention of transportation, people walked miles to survive, carving out new paths where none existed. Today, many simply travel the well-worn road without realizing they have the power to forge their own path.

**"When you stop working, you start to die!" - Ferruccio Lamborghini**

# Chapter 15:
# Forging Life

I have a friend whose son is a master swordsmith, skilled in the ancient art of folding metal. Whenever I visit their workshop, I find myself captivated by the mesmerizing process. He ignites the furnace, and soon, the fiery interior glows a fierce white-hot, illuminating the surroundings with an otherworldly light. With deft hands, he retrieves the glowing metal, its surface shimmering and alive with heat, and lays it upon a sturdy anvil. The moment he strikes it with his hammer, a brilliant shower of sparks erupts, dancing through the air like tiny stars exploding in a burst of creativity and labor.

This intricate dance of metal and flame mirrors our own journeys through life. At times, we feel as though we're trapped in a furnace, surrounded by overwhelming heat and uncertainty, with no clear direction in sight. As the pressures mount, we are transformed, emerging from the crucible of our challenges sometimes scarred, but often stronger. In those moments when life feels dull and uninspiring, we realize that, much like the metal, we must endure the fires and undergo repeated shaping to carve out a renewed path, discovering our true potential in the process.

Embrace the thrill of your journey! If you want to transform your life, you must take bold steps forward. Just like metal is forged into a fearsome sword through intense heat and countless folds, your own greatness requires the same dedication and resilience. Don't wait for life to hand you a treasure map; adventure awaits those who dare to seek it! It's all too easy to blend into the comfort of your surroundings, but take a close look at your circle. The moment you recognize that it's time to carve out a new path will be a turning point that ignites your spirit. Strangely enough, many people live in their comfort zone until their spark fades and passion dulls. Don't let that be your story! Go forth and shine bright!

**"On any given day, in any given moment, something could happen that opens up an entirely new path for you. Be still, be alert, be ready." -Marianne Williamson**

When my adopted father was nearing the end of his battle with cancer, I found myself sitting with him, reflecting on the complex and often turbulent relationship we had. In a moment of vulnerability, he confessed to me, "I never knew how to be a father." After years of silence between us, I looked into his eyes and replied, "Dad, it's not that hard. You have to care about your kids." I could see the weight of my words hit him hard; it was a moment of raw pain for both of us, as he seemed unprepared for that truth.

In that silence, I felt a deep sadness for him. I had forgiven my father long ago, but I also knew it was time for me to move forward with my life. I understood that as men, we often face a journey filled with challenges and disappointments. In those final moments together, I realized he had missed countless opportunities to truly be a father and a grandfather, trapped by his own beliefs and limitations. Our paths had been misaligned, and we were unable to navigate toward a genuine connection.

As I sat there, knowing he would soon be gone, I felt the weight of the name I carried his name. It was a powerful moment that made me resolve to ensure my own daughters would never feel the distance I experienced. I wanted his name to hold significance, not just for us, but in a larger sense, so I set my sights on achieving something remarkable: earning a place among the top ranks in the Army.

Those final days with my father were profoundly transformative for me. I came to understand the importance of redefining my own identity. He often shared a poignant story from his youth as an airman in the Air Force, filled with hope yet overshadowed by discrimination. That experience left a lasting impact on him, leaving wounds that never quite healed. As we navigate through life, we face a choice: to cling to resentment or to rise above and carve out a path toward our dreams. Each small step we take can lead us to a brighter, more fulfilling future, reminding us that we have the power to shape our own journeys.

"Anyone can steer the ship, but it takes a leader to chart the course. Leaders who are good navigators are capable of taking their people just about anywhere." -John C. Maxwell

# Chapter 16:
# Global Positioning Satellite

I vividly recall my first experience with land navigation in the Boy Scouts it was nothing short of a thrilling challenge! Growing up, I didn't have parents or family members who were into the great outdoors, so this was completely new territory for me. Land navigation is a fascinating blend of understanding the terrain, reading the weather, and honing your instincts. While we all chart our paths in life, it's surprising how many choose routes that lead them astray.

Diving into this unfamiliar skill set felt particularly significant for someone like me, as many children of color don't often step into environments like this. It was a transformative journey; land navigation not only taught me practical skills but also imparted vital lessons about what "true north" means in life. When you connect the concept of true north to your personal journey, something remarkable happens you begin to shift your mindset.

As I explored various terrains, it became clear: we all encounter both smooth and rocky pathways in life. The true north of existence lies in refusing to settle for a mundane life. I learned that to thrive, I had to embrace my uniqueness and become something extraordinary in my own eyes.

Life often presents us with formidable challenges, resembling a rugged path strewn with obstacles rather than the comfort of solid ground. Many individuals find themselves traversing these tumultuous roads for years, stuck in a cycle that drains their motivation to transform their circumstances. A significant contributor to failure lies in the inability to recognize when it's time to step away from a detrimental environment.

In the harsh landscapes of Iraq, survival meant finding solace in the simplest of pleasures like the cool embrace of fresh water and the reprieve of shade from the blistering sun. The necessity of wearing LPCs (Leather Personnel Carriers) became apparent, acting as vital footwear for navigating the unforgiving terrain, especially under the relentless heat. I quickly learned to tread carefully, avoiding the deceptive crunch of gravel and the treacherous softness of sandy surfaces to protect my ankles and knees from injury.

Understanding land navigation in life is crucial, as having a well-oriented life map can significantly impact your journey. If your life is not aligned with the right direction, you may feel lost and face challenges along the way. It's worth considering how often you may have had to reassess and redirect your path because of misaligned goals or values. Each day presents an opportunity to realign and set your course in the right direction.

Every individual, regardless of their background, must navigate their own journey, which may involve tackling various obstacles. While the experiences of someone born into affluence may differ from those of someone from a blue-collar family, both can encounter difficulties if they stray off course. The reality is that everyone will inevitably face challenges in life it's an inherent part of the human experience. Many people reflect on their lives by acknowledging that their choices have led them to their current situation, underscoring the importance of making informed decisions along the way.

In my final assignment in the Army, I faced a critical choice: enter the General Officer's next assignment pool or take the bold step of retirement. After a rewarding command experience and a challenging staff role during the COVID-19 pandemic, I made a decisive move for my future. I chose to retire and forge a new path in my life. It's essential to remain open to opportunities that exceed one's current rank or position. Reflecting on my father's experiences, where he confronted discouragement from others, only strengthened my resolve to carve my own path. My interactions with him prepared me for the challenges ahead. I vividly remember my last conversation with a senior commander where I asserted my decision by saying, "With all due respect, no thank you." For over three decades, I have dedicated my life to serving others, always upholding my dignity and honoring the sacrifices made by countless individuals of color who paved the way before me. Deep down, I recognize the weight of their sacrifices, which allowed me to stand on a more equitable playing field.

Those who came before me forged a path through discrimination, confirmation bias, and microaggressions. Each day, I lived for millions who never had the chance to stand up for themselves, enduring hardships without a voice. This was my 'General Colin Powell' moment. Legendary status is not achieved by chance: it arises from a crucial moment when you refuse to accept the status quo and take a stand against the behavior of others. Sometimes, you must sacrifice the moment to live with integrity. You need to look in the mirror each day and smile.

In my journey, I have become everything I have pursued. Each day, I find myself exactly where I am meant to be, thanks to that pivotal moment with my father that guided me to my true north. The ability to walk, see, and dream is a blessing that many do not share, and I embrace it wholeheartedly.

**"However beautiful the strategy, you should occasionally look at the results." -Sir Winston Churchill**

I don't depend on others for motivation. I drive myself. At some point in your life, the chips are all in your hands; the problem is, your enemies don't think you will use them. The moment I stopped caring about other people's perspectives, I gained the ability to make decisions that were in my best interest.

I was deeply moved by the film "Red Tails," particularly a moment that struck me with clarity. It made me realize that no matter how much effort I put in, I might never enjoy the same privileges as some of my peers.

In the movie, Colonel Bullard, the commanding officer of the legendary Red Tails, confidently declares that he remains unconcerned with the perceptions of non-colored officers regarding the African American Airmen under his command. His strong stance reflects both his unwavering leadership and the broader struggle for respect and recognition faced by his pilots during a time of racial segregation. Bullard emphasizes the importance of focusing on their mission and accomplishments rather than being distracted by prejudice, reinforcing his commitment to fostering a sense of pride and unity among his squadron. He knows they have the skills to excel if given a fair chance. Life would be simpler if everyone received equal opportunities, but that's not the reality we face. Recognizing that life demands transformation and the courage to adjust your path is crucial; otherwise, you risk settling for mediocrity.

The most exhilarating part of life is embracing the courage to change direction! It's not merely a one-time decision; it's a continuous adventure that requires regular updates, just like a cutting-edge GPS system. I learned early on that the only limits I face are the ones I impose on myself. I challenge myself to constantly identify what or who is standing in the way of my goals. I have immense respect for those pursuing opportunities after all, those who shy away from risk often don't add much value to any organization. Many people dream of success but hesitate to put in the hard work because they fear altering their circumstances. But here's the exciting truth: if you're ready for change, you can seize control of your life! Embracing your authentic self ignites a powerful transformation. Often, it's our past memories that keep us from discovering our true purpose. So, let's break free and unleash our potential!

**"It is the set of the sails, not the direction of the wind that determines which way we will go." -Jim Rohn**

Our life perspective clearly defines which way our GPS will take us. We frame our lives around our perspectives. The quality of your life hangs on your ability to reframe your life from a negative perspective. What do you want to be in life? Most people don't know what they want because they have no idea what direction to travel. The most powerful weapon in the world is how you see yourself. At some point, you have to navigate your life, and the key is deciding that you will no longer sit in still waters. You cannot sail out of troubled waters if there is no wind blowing. No one is an expert at the beginning of their journey.

**"Incredible change happens in your life when you decide to take control of what you do have power over, instead of craving control over what you don't." -Anonymous**

Life will inevitably throw a myriad of obstacles in your path, each one designed to veer your ship off its intended course. Yet, it is your internal compass a carefully calibrated mix of values, beliefs, and aspirations that will guide you through even the wildest tempests. This pivotal moment is where you seize control and actively define the narrative of your life.

When I reached the esteemed rank of Lieutenant Colonel (LTC) in the military, I took a moment to reflect and made a conscious choice to transition my life from the autopilot of complacency to a deliberate navigation of purpose and intention. I realized that simply going through the motions would not fulfill my potential. Determined to rise above the mediocrity that sometimes surrounded me, I embraced full accountability for my own decisions, understanding that each choice I made would ripple through both my professional and personal life. It was a moment of awakening one that ignited a passion for leadership, inspired growth in those around me, and set me on a course toward a more impactful existence.

The moment I grasped the steering wheel, I charted a course for success. My inner GPS illuminated a path dedicated to uplifting others, revealing that true fulfillment lies in enriching lives beyond my own. Your existence is a masterpiece waiting to be created, so navigate your vessel toward tranquil waters. Remember, mastering your emotions is essential for steering your destination.

This brings me to the next compelling chapter of my journey a chapter focused on an extraordinary young man named Jake Derrico, whose example will forever inspire me to strive for excellence.

# Chapter 17:
# The Derrico Effect

**"Things we lose have a way of coming back to us in the end, if not always in the way we expect." -J.K. Rowling**

On a bitterly cold winter night in Wausau, Wisconsin, during the late 90s, the sky loomed overhead, heavy and shrouded in darkness, as if mourning the world below. I settled into the familiar embrace of my living room, cocooned in warmth, when suddenly, the piercing ring of my phone shattered the stillness, cutting through the quiet like a bolt of lightning.

The voice on the other end delivered news that landed with the weight of a thunderclap: a soldier's family had just received the devastating news of their son Jake's untimely death in a car accident. My heart raced as I absorbed the gravity of the situation I had only 24 hours to make my way to Curtiss, Wisconsin, assigned as the Casualty Assistance Officer. The enormity of this solemn responsibility pressed down on me, a young Captain adorned with less than six years of service in the Army, and I felt the full weight of what lay ahead.

Steeling myself for what lay ahead, the next morning, I stepped into the early morning as it was still dark, where the wind howled like a mournful spirit, and the air bit like shards of ice against my skin. The roads were a treacherous sheet of black ice, glimmering ominously in the moonlight, making every turn feel perilous. As I navigated the desolate landscape, an unsettling thought loomed: what could I possibly say to the family in their profound grief? I grappled with the reality that in moments of such heart-wrenching loss, words often felt woefully inadequate. As a father myself, I understood the unnatural agony of parents burying their children an event that no one is truly prepared to endure.

The Derricos were a resilient family, steadfast and deeply intertwined with the soil they had cultivated for generations. Jake's father, Frank, embodied the essence of a Marlboro man rugged, strong, and sun-kissed, a living testament to decades of toil beneath the wide expanse of sky. His mother, Judy, radiated an effortless beauty, her striking features etched with the spirit and tenacity that define Central Wisconsin, her hands adept at nurturing both the earth and her family through the traditions of blue-collar life.

As I approached their home, a biting cold gnawed at my bones, seeping into my very core. The air was thick with anticipation as I knocked on the weathered door, which swung open to reveal Frank and Judy. They stood there, visibly shaken yet somehow maintaining an air of composure, their expressions a blend of worry and resolve. As I clasped Frank's hand, it felt as unyielding as granite a man who had faced countless adversities with unwavering courage throughout his life.

Stepping inside, the atmosphere was thick with sorrow, yet the presence of family and friends gathered in their home seemed to form a protective cocoon around them. My heart raced with nerves; the weight of the moment sat heavily on my chest. Yet, in that sea of grief, it was Judy who broke through my anxiety. She approached me, her embrace warm and genuine, whispering, "Thank you for coming."

Once I stepped into the kitchen, I felt an overwhelming weight settle in my chest, the enormity of the situation leaving me momentarily speechless. "Could you share a picture of Jake with me?" I asked softly, my voice barely above a whisper, hoping to spark a glimmer of warmth amid the pervasive sorrow. As she began to recount cherished memories, the stories flowed like a gentle river each tale illuminating Jake's vibrant spirit as a boy and a young man. Laughter intertwined with tears, creating a poignant melody as everyone in the room joined in the reminiscence; what started as an almost suffocating sadness slowly transformed into a heartfelt celebration of a life deeply loved.

It was at Jake's funeral, amid the tear-streaked faces and whispered recollections, that I would truly transform. With each step away from that hallowed ground, I would carry the heavy yet honorable burden of their loss, forever altered by their unwavering strength and the legacy of their beloved son.

After several days of paperwork and meetings with the family, the time had come for the funeral. As his Platoon Sergeant delivered Jake's Eulogy, I knew this moment would transform my life forever. The Platoon Sergeant recounted how Jake single-handedly shifted his unit's mindset, inspiring everyone around him.

Jake, a junior enlisted soldier in an infantry unit at Fort Carson, Colorado, had a burning ambition to become an Airborne Ranger. As he integrated into his unit, he noticed that not all of his fellow soldiers shared his dedication to training. Determined to make a positive impact in the Army, Jake focused on his goal of achieving the prestigious Ranger designation. With a profound appreciation for his country and inspired by his grandfather's military service, he remained steadfast in his pursuit of excellence.

During peacetime, military units undergo rigorous assessments of their operational readiness through a series of comprehensive training deployments, which encompass both non-kinetic and kinetic operations. Jake's unit was slated to deploy to the Joint Readiness Training Center at Fort Polk (now known as Fort Johnson) in Louisiana. It quickly became apparent, through the observations of his Platoon Sergeant, that Jake was not just an ordinary soldier; his remarkable skills and unwavering dedication had the potential to profoundly elevate the unit's overall performance during the demanding exercises at JRTC.

On a balmy afternoon at Fort Carson, the unit was deeply immersed in training exercises, diligently preparing for the upcoming Joint Readiness Training Center (JRTC) deployment. As the sun began to set, casting a golden glow across the grounds, the Platoon Sergeant gathered the soldiers in a tight circle, the murmur of conversations fading into an attentive silence. He addressed the group, probing whether everyone felt all tasks had been thoroughly accomplished. With a warm yet authoritative tone, he encouraged anyone who felt that more training was necessary to raise their hand.

After a moment of hesitant contemplation, Jake, a dedicated soldier known for his unwavering determination, slowly raised his hand. His voice rang out clear and resolute as he stated, "I believe we need to train a little bit longer." The Platoon Sergeant paused, his brow furrowing slightly as he weighed Jake's suggestion against the collective needs of the platoon. The decision to extend the training sessions over the ensuing days was made, setting the stage for intense days ahead.

As the week unfolded, however, the atmosphere shifted. Some members of the platoon began to voice their growing discontent regarding the longer training hours, their fatigue palpable. In an effort to understand the driving force behind Jake's advocate for additional training, the Platoon Sergeant called him into his office.

Sitting across from each other, the two engaged in a candid conversation that illuminated Jake's motivations. He articulated his belief in striving for excellence, aware of how his perseverance was impacting both himself and the morale of his peers. These discussions served as a pivotal moment, revealing not only Jake's commitment but also hinting at the remarkable transformations that lay ahead for the entire unit. The seeds of progress were being planted, promising noteworthy developments in their journey together.

**"You just can't beat the person who never gives up." -Babe Ruth**

The discussion between the Platoon Sergeant and Jake highlights the importance of effective training and personal growth in competitive environments. When the Platoon Sergeant inquired about Jake's concerns with the current training regimen, Jake respectfully shared his background as a former top karate fighter in Wisconsin. He emphasized that achieving excellence in competition requires not only skill but also the right training approach.

Jake vividly recounted his experience during his inaugural year in high-level competition, a journey marked by fierce determination and disappointment. In the finals, he faced a challenging opponent and ultimately fell short, a bittersweet moment that lingered in his memory. A seasoned karate master, impressed by Jake's technical prowess, delivered a piece of candor that struck a deep chord within him: while his skills were commendable, he lacked the instinctual finesse that could elevate him in the dojo.

This invaluable feedback ignited a spark of introspection in Jake, compelling him to reassess his training approach. Understanding the need to cultivate an instinctual responsiveness to the unpredictable nature of a fight, he set out on a rigorous path of self-improvement.

With unwavering resolve, Jake immersed himself in a year of relentless training, his days filled with grueling drills and sparring sessions that tested his limits. As he honed his techniques, he began to weave them into a fluid tapestry of motion, where each movement flowed seamlessly into the next, a dance born from both discipline and instinct.

His relentless dedication soon bore fruit. When he returned to the tournament the following year now brimming with confidence and honed instincts he not only faced his opponents but danced with them, anticipating their every move. The culmination of his hard work and evolution in skill culminated in a triumphant victory, a moment that transformed his initial setback into a powerful testimony of resilience and growth.

By sharing his journey with the Platoon Sergeant, Jake helped provide insight into his commitment to training and his vision for improvement. This exchange motivated the entire platoon, leading to a shift in their training mentality. As a result, during their subsequent deployment to the Joint Readiness Training Center (JRTC) that Spring, the unit surpassed their performance benchmarks, largely attributed to Jake's remarkable work ethic and determination.

**"Remember that guy that gave up? Neither does no one else." - Anonymous**

As I sat in my chair listening to the Platoon Sergeant's speech, I took in the farm where Jake had grown up. His incredible work ethic was forged in the semi-arctic environment of Central Wisconsin. Born in God's country, Jake embodied manners and selfless service. He once placed a knife in a tree and told his father that he would remove it once he completed Ranger school. Tragically, that knife would remain in the tree forever. When the honor guard fired its 21-gun salute, I was overwhelmed with emotion. Upon entering the house, Jake's grandfather approached me with tears in his eyes, asking me to make a promise in Jake's honor.

This request would become the cornerstone of my military career, shaping my leadership philosophy and approach to service. I pledged to rigorously train and prepare my soldiers in Jake's memory, honoring him from that moment onward. Every opportunity to share Jake's story his bravery in the face of danger, his unwavering commitment to his comrades, and the ultimate sacrifice he made became a chance to inspire my troops to strive for greatness and embody the values he represented. I committed myself to ensuring that my units were not only equipped for combat but also mentally and emotionally prepared for the challenges of war, unaware that a protracted twenty-year conflict awaited us on the horizon.

Though Jake's story had a heartbreaking conclusion, its impact was profound on all it touched, motivating countless individuals to embrace the ideals of courage and resilience. The Derrico Effect encapsulates the essence of selfless service and the readiness to confront life's inevitable challenges, reminding us that our actions can leave an indelible mark on the world around us.

**"It always seems impossible until it's done." -Nelson Mandela**

# Chapter 18:
# Keep It Rolling

**"Either write something worth reading or do something worth writing." -Benjamin Franklin**

"Keep It Rolling" is all about unleashing your passion and transforming your life! You must find your own way; it's a thrilling journey. Throughout my life, naysayers have tried to limit me, insisting on what I can't achieve. Then came 2020, a year that turned the world upside down during COVID-19. For the first time since the Great Depression, the wealthy faced challenges, while those once downtrodden found a level playing field. The pandemic shone a light on the illusion of those who were "fake rich," compelling everyone to face their reality. Amid the chaos, I reassessed my life, just as I did during combat. The magic begins when you break free from cruise-control living!

Life can be viewed as a complex journey filled with various intersections, each presenting opportunities for growth and exploration, particularly when we reflect on our successes and setbacks. It's important to recognize the significant roles that mentors such as veterans and educators play in shaping our paths. For example, individuals who enlist in the military often encounter challenges such as racism, sexism, and gender bias. Research and personal accounts indicate that these issues persist in military environments and can affect individuals regardless of gender. It is crucial to acknowledge and address these forms of discrimination in the workplace to ensure a fair and inclusive atmosphere for everyone.

We all must navigate the complexities of society. Hard times can linger if you surrender to chaos, but we must forge ahead because quitting leads to disaster. Reflect on the lies you've told yourself to fit in with people who genuinely don't care about you. While wrestling with various challenges, I met an incredible guy named Dave.

Dave and I forged a strong friendship at the Command and General Staff College, and one day, I noticed his wallet was open, revealing a food stamp. Curiosity got the better of me, and I asked this Army Officer why he carried it. He explained that it serves as a constant reminder to stay humble. Although his family once depended on food stamps, Dave made the courageous decision to return to college and become an officer. Holding onto his last food stamp symbolizes his unwavering determination to remember his roots. He emphasized that many military officers remain unaware of the struggles others face. Life may often seem like a series of choices, especially for those who have had a smoother path. Yet, when adversity strikes, it's often those who have never known hardship who seek compassion the most.

**"Focus on what lights a fire inside of you and use that passion to fill a white space. Don't be afraid of the challenges, the missteps, and the setbacks along the way. What matters is that you keep going." - Kendra Scott**

Never let excuses overshadow your opportunities. I once met a young officer who was struggling with a profound sense of inadequacy, convinced that his voice was lost amid the noise of the ranks. Sensing his potential, I encouraged him to share his ideas. As he spoke, I was captivated by the fervor and dedication radiating from him, a true desire to foster positive change within the Army.

After our conversation, I made a pivotal decision: I transferred him to my unit and entrusted him with a command. It turned out to be one of the best choices I ever made. He transformed into one of the most effective commanders I have had the privilege to work with. By reshaping his journey, I inadvertently reignited his passion for the Army a flame that continues to burn brightly to this day.

We must always commit ourselves to the pursuit of making the world a better place, recognizing that sometimes all it takes is a moment of encouragement to unlock someone's true potential.

**"To live is the rarest thing in the world. Most people exist, that is all." -Oscar Wilde**

When the sun shines on my tombstone, I want it to embody my passion for making a difference in the world. Many believe that completing one significant task marks the end, but in reality, it's only the beginning. We all have 24 hours each day; it's how we use them that sets us apart. We must cherish every moment that God grants us.

Just like me, anyone can overcome their past. We all encounter depression in the terrain of our lives, and it's vital to learn how to cope with those moments. True transformation to greatness is not measured by money, but by the lives we touch. Ask yourself: are you really living, or just getting by? We will all face setbacks, but we have the power to change our life's mission.

A memorable experience I had involved a man who kindly paid for my meal, motivated by a desire to "pay it forward and make the world a better place." This encounter inspired me to adopt a similar approach. I now make a conscious effort to cover meals for single parents or blue-collar workers whenever I can, believing that small acts of kindness can spark hope and reinforce the possibility of positive change. Even the simplest gestures can significantly brighten someone's day, and I strive to be a kind of "angel" for those I meet along the way, even if they are strangers. Ultimately, my goal is to leave a lasting legacy through my acts of kindness.

The philosophy of "Keep It Rolling" has the power to transform one's life for the better. Each person encounters unique challenges that may seem overwhelming. By bravely confronting these obstacles instead of shying away, individuals can unlock the door to meaningful and lasting progress. Life offers a rich tapestry of experiences filled with invaluable lessons, inviting us to reassess our paths and aspirations.

Happiness is not just a stroke of luck or a fleeting feeling; it is an intentional commitment that we each must choose to embrace. The journey toward happiness is deeply personal, calling us to actively nurture our joy and sense of fulfillment.

To embark on this path, we must navigate our emotional burdens with care and attention. Managing our feelings and reactions is crucial, as adopting a victim mentality can hinder our personal growth. This involves recognizing negative thought patterns and consciously shifting our mindset toward positivity. By committing to our well-being and acknowledging our role in shaping our happiness, we open ourselves to a richer, more fulfilling life.

Each individual embarks on a unique journey defined by their own mission, strategy, and destination. By understanding and skillfully navigating the intersection of these elements where purpose aligns with plans and goals we unlock the door to a more extraordinary life experience. Embracing this concept empowers us to Keep It Rolling, moving forward with resilience and optimism as we pursue our remarkable paths.

# Gallery

**Georgia Southern ROTC Summer Camp**

**Georgia Southern Commissioning Day,
Mrs. Helen Jackie Yates  (Mama Yates)**

**Iraq 2007 Coalition Team**

**Iraq 2007 Coalition Officer**

**Iraq 2004 Support Ops Officer**

**Iraq 2004 Al ASAD**

**Iraq 2004 Al ASAD**

**Iraq 2004 Joint Chief of El Salvador**

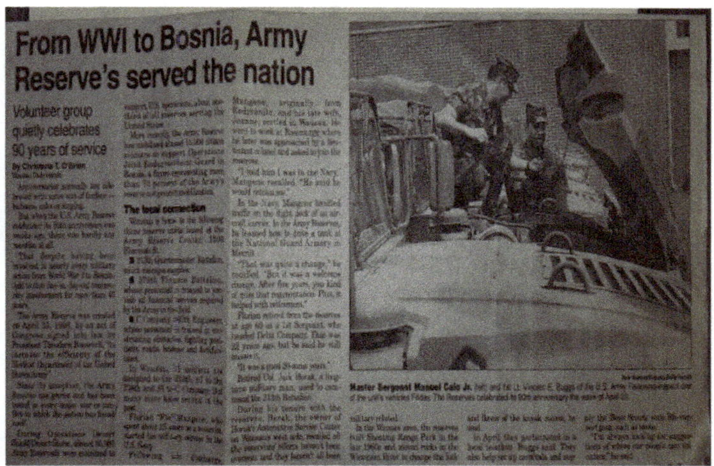

**Wausau, WI 1997**

**Ret Master Sergeant Air Force AB Spencer (Stepfather)
Air War College Graduation**

**Deployment BBQ**

**Promotion to COL**

**90s Training Exercise**

**The Georgia Southern Eagle Flying High after a Base Attack**

**First Day in Kuwait before we Push North**

**Awarded the Polish Bronze Medal**

**2004 Al Asad, Iraq**

**Retiring My Mother General Mom   Call Sign GM**

**Loving Father, ret. Master Segreant
Alton G. Buggs now Rest at  Fort Sam Houston**

**Great Command Team 364th ESC**

**Fort Polk, LA Joint Readiness Training Center**

**Enjoying Europe with Cheryl**

# Soldier who became pen pals with kindergartners 13 years ago finally meets them

October 29, 2019 by FarahR

**Story Coming Soon in my next Project**

Members of the K-5 class at David Emanuel Academy are shown holding small American flags and certificates authenticating that these flags were flown over Iraqi headquarters. Currently, the students are corresponding with a soldier stationed in Iraq. He then obtained certificates with each student's name on them, stating that the flags were flown in their honor. Since then, the kindergarten class has gathered a care package of food and GSU items (the Major is a GSU graduate and a fan of GSU athletics).

**Story Coming Soon in my next Project**

# Conclusion

98

Each of us is called to a unique purpose that fuels our aspirations. The primary aim of this book is to help you uncover your personal mission, identify your destination, and devise a powerful strategy to achieve it. To find your true north the guiding principle that aligns with your values and passions you must engage in self-reflection and surround yourself with a positive and supportive network. Remember, no one reaches the pinnacle of success without the encouragement and guidance of the right mentors.

"Keep It Rolling" highlights the importance of proactively seeking opportunities and thriving in the face of challenges. It's about not only recognizing potential paths for growth but also equipping yourself with the skills and mindset to overcome obstacles. By cultivating resilience and adaptability, you can excel and continue to move forward, regardless of the circumstances.

# **Acknowledgments**

I want to give a heartfelt shoutout to everyone who played a vital role in completing this project. Special thanks to my editing team, Mr. Adam Musgrave from Newsome High School, who tackled the initial challenging draft, and Doug Thurston, who polished the final edits. I'm deeply grateful to my podcast team, Dr. Jennifer Blalock and Mike G, for always having my back while I focused on this project. My thanks also go to part of my Embrace the Power staff, Kelly Hoing and Mary McCarty, for your unwavering support, and to my Web Service team, Marina Griffin.

Lastly, a huge thank you to my inspiration team my true friends who lifted me up and encouraged me to keep going: my fitness trainer, Juan Musibay, and football coaches Eric Brown, Glenn Castle, and Bruce Gifford. Your friendship has been a lifeline. To my extended blended family, Alton Buggs II and Ryan Buggs, keep rolling, my little brothers!

# About the Author

**Ret. Brigadier General, United States Army**

## VINCENT E. BUGGS

BG Buggs's journey into the military realm was profoundly influenced by his upbringing as a military dependent. Growing up in this environment, he gained a profound understanding of the sacrifices and challenges faced by military families. This background has fueled his deep appreciation for the relationships and camaraderie that are cultivated within these close-knit communities, further motivating him to strengthen these bonds throughout his career.

BG Buggs is a remarkable leader who actively promotes personal growth and cultivates a success-driven mindset among those he mentors. His impressive academic background significantly amplifies his inspirational journey. He earned a master's degree in Global History from the American Military University, where he gained a nuanced understanding of historical dynamics that inform strategic decision-making. This foundation is further strengthened by his master's in strategic studies from the U.S. Air Force Air War College, a prestigious institution known for developing future leaders in national defense and military strategy.

BG Buggs' educational journey began with a bachelor's degree in history, complemented by a Minor in German from Georgia Southern University, which he completed in 1990. This degree not only reflects his academic prowess but also underscores his dedication to continuous learning and cultural understanding. Beyond his academic qualifications, BG Buggs possesses extensive professional experience that allows him to transcend the typical mentor role. He embodies a transformative presence, empowering individuals to identify their strengths and cultivate their potential. By instilling confidence and providing actionable strategies, he enables them to achieve meaningful success in both their personal and professional pursuits. His commitment to mentorship and growth makes a lasting impact on everyone he encounters.

## ASPIRE TO INSPIRE

Brigadier General (BG) Buggs has built an illustrious career, highlighted by a series of prestigious awards that underscore his exemplary service and leadership. He has been honored with the Legion of Merit, a testament to his exceptional commitment and achievements in the military, along with Joint Service Awards recognizing his collaborative efforts across various branches of the armed forces. Among his notable distinctions are the Combat Action Badge, the Distinguished Star from El Salvador, awarded in recognition of his impactful contributions during international missions, and the Polish Bronze Medal, celebrating his role in fostering strong international military partnerships.

In addition to his military accomplishments, BG Buggs has taken on the roles of coach, mentor, and leader with a passionate dedication that spans more than 40 years. His unwavering commitment to developing the potential of those around him has left an indelible mark on countless individuals, inspiring them to strive for excellence in their own careers.